TRACE ANALYSIS OF ATMOSPHERIC SAMPLES

TRACE ANALYSIS OF ATMOSPHERIC SAMPLES

Kikuo OIKAWA
Niigata College of Pharmacy
Hokuto College of Technology
Kamishin'ei-cho, Niigate 950–21, Japan

A HALSTED PRESS BOOK

KODANSHA LTD.
Tokyo

JOHN WILEY & SONS
New York–London–Sydney–Toronto

KODANSHA SCIENTIFIC BOOKS

Library of Congress Cataloging in Publication Data

Oikawa, kikuo.
Trace analysis of atmospheric samples.

"A Halsted Press book."
Includes index.
1. Air--Analysis. 2. Trace elements--Analysis.
I. Title.
QD121.038 628.5'3 77-3458
ISBN 0-470-99013-9

Published in Japan by
KODANSHA LTD.
12–21 Otowa 2-chome, Bunkyo-ku, Tokyo 112, Japan

Published by
HALSTED PRESS
a Division of John Wiley & Sons, Inc.
605 Third Avenue, New York, N.Y. 10016, U.S.A.

PRINTED IN JAPAN

Preface

In recent years, the sensitivity and precision of analytical instruments have been markedly improved, and progressive modifications towards automatic operation have been made. Thus, most manipulatory factors contributing to analytical errors have been eliminated, although the collection and pretreatment of samples still depend largely upon human manipulation. Needless to say, proper analysis and assessment of environmental pollutants depend upon accurate survey results.

In this book, methods for sample collection and pretreatment leading to analysis of the metal content of atmospheric particulate matter are described, together with analytical procedures. Chapter 2 covers the fundamentals of atmospheric particulate sampling, high volume air samplers, low volume air samplers, and the Andersen sampler, placing emphasis on flow rate calibration and filter choice for proper sampling. Chapter 3 describes the most suitable pretreatment methods for applying the chosen analytical method to the collected sample, and discusses problems such as contamination arising at the pretreatment stage. Chapter 4 covers the analytical methods, including atomic absorption spectroscopy, emission spectroscopic analysis, X-ray fluorescence analysis, X-ray diffraction, and neutron activation analysis.

The text deals mainly with the most recent methods for pretreatment and analysis, and is intended primarily for engineers actively engaged in work in this field.

The author is indebted to the staff of Kodansha for their editorial and linguistic assistance in the preparation of this book.

February, 1977 Kikuo OIKAWA

Contents

CHAPTER 1

INTRODUCTION

Since the materials causing air pollution, that is "suspended particulate matter" or "suspended dust", have various chemical and physical properties, their effects on human health and the environment are not uniform, and these effects are further enhanced by the presence of various gaseous materials.

1.1. Physical properties

The suspended particulate matter consists of particles of various sizes, ramging from single molecules (0.02 μm in diameter) to large particles (500 μm in diameter), which may be in the solid or liquid state, and remain in the air for several seconds to several months depending on their size. Particulates larger than 1 μm show a high sedimentation velocity and move independently of the wind, while particulates smaller than 1 μm show a low sedimentation velocity and move with the wind. The movement of particulates smaller than 0.1 μm resembles that of a gas and is possibly identical to diffusion. Such small particulates easily adsorb gases, collide with each other, and undergo adhesion. It is probable therefore that particulates smaller than 1 μm exert a stronger influence on human health and the environment.

The relationship between the diameter of the particulate material and its precipitation in the lung alveoli will next be considered. The respiratory air passages begin at the nasal cavity, pass through the trachea, bronchi and bronchioles, and end at the alveoli. The diameter of these passages and the velocity of air flow through them also decrease in this order. Larger particulates are therefore unable to penetrate into the narrow branches of passages, but smaller ones can easily reach the alveoli. Particulates larger than 10 μm are mostly trapped in the nasal cavity and throat, while 90% of the particulates larger than 5 μm precipitates in the upper part of the air passages. Particulates in the 1–5 μm size range precipitate in the trachea and bronchi. Fine particulates smaller than 1 μ reach the bronchioles and alveoli. However, the rate of precipitation of particulates in the 5–0.5 μm range decreases in proportion to diameter, the precipitation rate of particulates of 0.5 μm diameter being about 20–30%. Much smaller particulates then precipitate at a greater ratio. In the alveoli, maximum precipitation is observed with particulates in the 1–4 μm size range, and the minimum is with those of 0.4 μ diameter. Fine particulates smaller than 0.4 μm again precipitate at an increased rate.

Clearly, therefore, it is important to investigate the size distribution of particulates in addition to simple quantitative observations on their total weight.

1.2. Chemical properties

The inorganic components of suspended particulate matter originate from the soil, earth (volcanic activity), and other artificial routes, and comprise oxides, salts, organic chelates and other compounds of various types including almost all elements. In areas adjacent to active volcanoes, higher concentrations of heavy metals such as mercury, arsenic, selenium, lead, and vanadium have been detected, while mercury and arsenic (in addition to hydrogen sulfide) are often detected at high concentrations in the vicinity of subterranean geothermal power plants. Of course, the particular concentrations and composition of such contaminants depend very much on the local conditions.

Great varieties of inorganic pollutants are produced artificially. Vanadium, nickel, selenium and mercury are discharged as exhaust fumes from oil refineries and boilers using heavy oil or coal. The exhaust fumes of copper and zinc refineries represent a source of zinc and cadmium. Iron, manganese, zinc and arsenic are discharged from ironworks. Difiusion of lead from cars occurs when tetraethyl lead is added to gasoline to improve its combustibility. On the other hand, in areas adjacent to tracks for electric trains, cadmium is often detected at high levels. It originates from the cadmium-alloy wire to which this metal is added to improve hardness.

Domestic wastes may be burnt or used for land reclamation. Incinerators of waste materials and sludge thus produce all kinds of metals. In particular, mercury concentrations may be high. The metal originates from spent fluorescent lamps and mercury cells disposed of in the office and home, and the source in sludge includes drain effluents carrying mercurochrome and other mercury-containing medicines used in hospitals. The drainage from hospitals is in fact one of the important sources of mercury pollution of rivers, canals and bays in urban districts, and more careful management is required to reduce the dangers to human health.

The pollution of lead from car exhausts now represents a major social problem. It arises from the tetraalkyl lead which is added to gasoline to elevate the octane number and is exhausted into the air in the combustion process. Since about 70% of the lead in the air is in the form of particulates smaller than 1 μm in diameter and 50% consists of particulates smaller than 0.5 μm, according to our investigations, these can easily peentrate into the alveoli. Although the type of lead in the exhaust gas depends to some extent on driving conditions, 30–90% is PbClBr and the remainder consists largely of complexes of PbClBr and NH_4Cl, e.g. α-$NH_4Cl \cdot 2Pb$ ClBr, β-$NH_4Cl \cdot 2PbClBr$, and $2NH_4Cl \cdot PbClBr$. The sources of Cl and Br are the ethylenedichloride and ethylenedibromide added to gasoline as lead-clearing agents. Small quantities of $PbSO_4$ and $PbO \cdot PbClBr \cdot H_2O$ are also detected, and $3Pb_3(PO_4)_2 \cdot 2PbClBr$ forms about 20% of the lead in car exhaust when phosphorus additive-containing gasoline is used. The major lead compounds in the combustion and exhaust chambers of engines are PbClBr and $PbO \cdot PbClBr \cdot H_2O$ and these probably react with small quantities of NH_3 in the exhaust gas to yield complexes of PbClBr and NH_4Cl.

In the U.S.A., in addition to lead pollution from car exhausts, impediments of infant health arising from the sucking, chewing and ingestion of lead-containing paint from walls and furniture has also become a problem.

In order to examine the influence of metals on human health and to explain the chemical mechanisms of air pollution by metals, it is important to obtain not only quantitative but also detailed qualitative information regarding each element. In the past, there has been a tendency for only the quantity of each element detected as metal to be discussed. However, experimental investigations with rats have indicated, for example, that the toxicity of $MnSO_4$ is about 100 times that of MnO_2. In general, soluble compounds of metals display a stronger effect on human health than insoluble ones. Vanadium, manganese, nickel and iron, acting as oxidative catalysts under conditions of high humidity, accelerate the formation of sulfate and nitrate mists from sulfur dioxide and nitrogen oxides, although their individual reactivities of course vary according to valency.

CHAPTER 2

SAMPLING

2.1. Fundamentals of sampling

A. Selection of suction pump

It is desirable that the suction pump provides a high vacuum, has a large flow volume, causes minimal pulsation, and is easily portable. However, all of these conditions are rarely satisfied. Several types exist among available suction equipment, e.g. rotary vacuum pumps, piston air pumps, diaphragm pumps, and blower motors. In this section, details of the rotary vacuum pump and blower motor will be given since these two are widely used for the sampling of suspended particulate matter.

a. Rotary vacuum pump

Pumps of the type generally known as rotary vacuum pumps are widely used for the sampling of particulate and gaseous matter. Their typical inner structure and external appearance are illustrated in Figs. 2.1 and 2.2, respectively. A rotoris mounted eccentrically within a casing, and the rotor is fitted with ditches in which sliding vanes are mounted. Four vanes are generally installed to reduce pulsation, although a single vane works

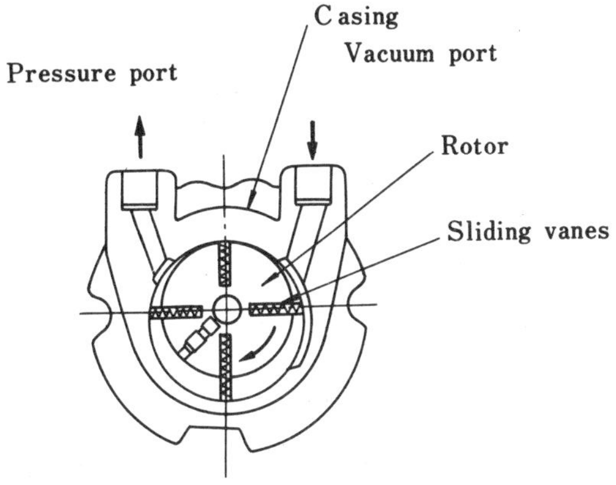

Fig. 2.1. Section through a rotary vacuum pump.

A

B

Fig. 2.2. Vacuum pumps. A, Rotary vacuum pump; B, diaphragm vacuum pump.

well. As the rotor moves, the vanes travel in and out along the ditches due to centrifugal force, and one and of each vane touches the inner wall of the casing. The crescent-shaped space enclosed between the rotor and casing is divided into small compartments, and the air is compressed. This process is illustrated schematically in Fig. 2.3.

Suitable equipment may be selected from various commercial products with the general properties listed in Table 2.1, depending on the purpose, required flow volume for routine work, and degree of vacuum. Pumps of with a power of 200 W and 400 W are commonly used as low volume air samplers (20–30 1/min) and middle volume air samplers (100 1/min), respectively. The vacuum gained with such pumps is a little lower (500–600 mm Hg) than that with an oil pump. Under reduced pressure, the suction volume is not equal to the exahust volume, so that it is incorrect to determine the flow volume at the pressure port of the pump.

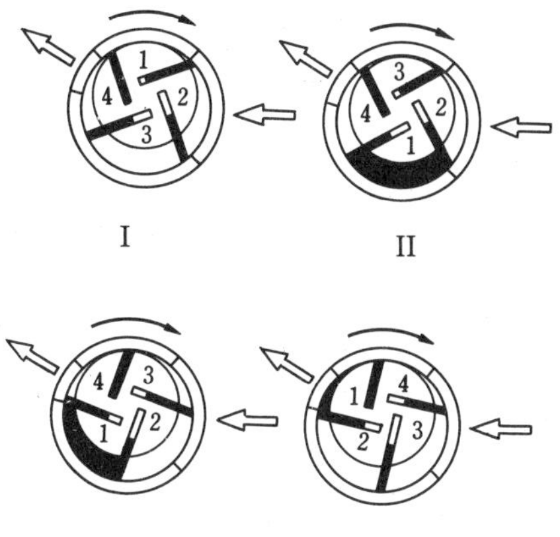

Fig. 2.3. Idealized cross section of a rotary vacuum pump in operation.

TABLE 2.1 Characteristics of some vacuum pumps of different power

Power	Compression characteristics			Weight (kg)
	Max. pressure (kg/cm³)	Continuous pressure (kg/cm³)	Outlet air flow at continuous pressure (l/min)	
20 W	0.1	0.1	{10 (50 Hz) 12 (60 Hz)	4
35 W	0.3	0.2	30	6.7
200 W (Maker H)	0.5	0.4	55	13.5
(Maker I)		0.5	{125 (50 Hz 145 (60 Hz)	
400 W (Maker H)	0.5	0.4	130	18
(Maker I)		0.5	{208 (50 Hz) 250 (60 Hz)	

In practice, the pump should be operated carefully and properly taking into account the following items.

(1) Setup site. A flat, dustless, rain-proofed, ventilated place should be chosen. Special care is necessary during the rainy season.

(2) Exchange of sliding vanes. The vanes must be renewed after 8000 h of operation (i.e. about 1 yr for non-stop operation).

(3) Cleaning of filter (vacuum port) and silencer (pressure port). These should be cleaned at least every 3 months and renewed every 6 months.

(4) No lubrication should be undertaken.

(5) The pump should be operated without load for about 30 min every 3 months if it is in continuous use for a long period.

(6) As a rule, the gas meter or flow meter should not be connected directly to the pressure port of the pump. Abrasion of the sliding vanes produces carbon dust which may contaminate the flow meter and consequently lead to errors in the volume indication.

b. Blower motor (fan motor)

The rotary vacuum pump described above and the diaphragm pump increase in weight in proportion to increase in their suction volume, so that their portability progressively decreases. In practice, the upper capacity limit of such pumps is about 100 l/min. On the other hand, blower motors can attain a suction volume of as much 500–1500 l/min. The blower motor is commonly used as a high volume air sampler. It consists of a double centrifugal turbine pump directly connected to a series motor of high speed, and attains a suction volume of 1.3–1.5 m^3/min. Although it is unsatisfactory for extended operation since the static pressure is low, it is highly useful for short-term operation such as sampling for 24 h. There is no problem with the use of glass fiber filters, but membrane filters cannot be used due to the excessive pressure reduction.

B. *Flow measurement*

The devices widely used for flow measurement are the float area flow meter, orifice flow meter, and gas meter. In this section, the first and third devices will be described in detail.

a. Area flow meters

There are several types of area flow meters, e.g. the float type, piston type, and gate type. Of these, the float type is the most widely used. The principle is as follows. A float is mounted in a vertical tapered tube. As fluid enters the lower end of the tube, its flow is obstructed by the float and a pressure difference arises between the front and rear of the float. The latter is then lifted up by the resultant upward force, and comes to rest at the point where the pressure is balanced by its effective weight. Since the distance of float movement, or current area, is correlated to the flow volume, the latter can be determined by reading off the position of the float (Fig. 2.4).

$$Q_0 = CA\sqrt{\frac{2gV_f}{A_f}\left(\frac{r_f}{r_0} - 1\right)}, \tag{2.1}$$

where Q_0 is the flow volume, C the flow coefficient, A the current area be-

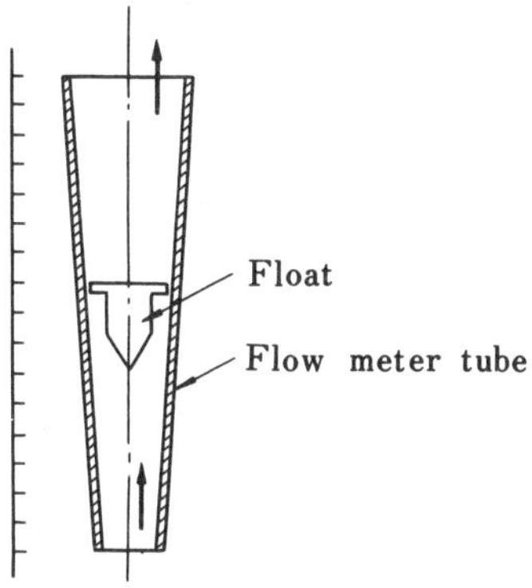

Fig. 2.4. Cross section of a rotameter.

tween the tapered tube and float, V_f the volume of the float, A_f the effective float area, r_f the relative float weight, r_0 the relative fluid weight, and g the acceleration of gravity.

A typical float area flow meter is the rotameter, in which the float side is ditched to rotate, so allowing easy balancing. Rotation of the float reduces friction with the fluid and lessens the influence of its viscosity, so that the float quickly balances. Meters with nonrotating floats are often called rotameters, but they are better termed Fischer-type rotameters. As shown in Fig. 2.5, the shape of the float may be modified depending on the fluid used, to improve stability and reduce the influence of the fluid viscosity. The floats are made of various materials such as Nylon, glass, aluminum, and SUS 27.

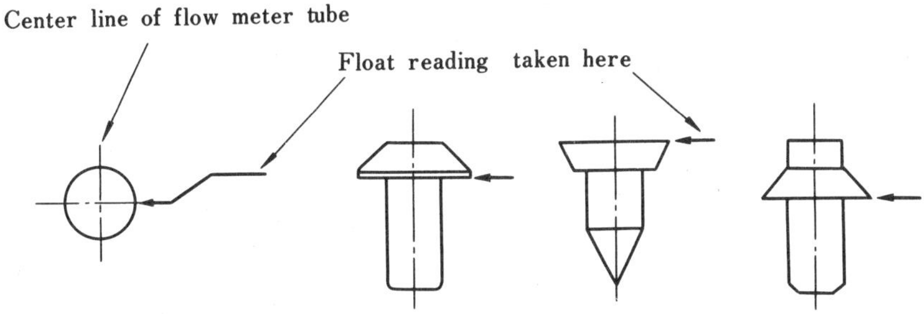

Fig. 2.5. Three kinds of rotameter floats. Readings are conventionally taken at the widest point of the float.

Some further details on the use of area flow meters will be given next. The rotameter is generally operated under increased pressure, but under reduced pressure in the case of air sampling. An example of the system employed is illustrated in Fig. 2.6. The dotted line shows the range of reduced pressure. The pressure conditions depend on the filter used, quantity of particulates stuck on the filter, the size of the filter, and the suction volume. Since the air density is low under reduced pressure, the flow meter gives a value several times to 30% higher than the true value. Fig. 2.7 shows plots of pressure change against face velocity for various filters. Clearly, the greatest pressure change is that caused by membrane filters. The pressure is not constant during the sampling operation: the pressure difference increases according to the quantity of particulates stucked on the filter and is also affected by temperature. The value for the flow volume given on the meter cannot therefore be corrected to that under standard conditions (say, 20°C and 760 mm Hg) by any general equation. Several methods exist to solve such problems, as follows.

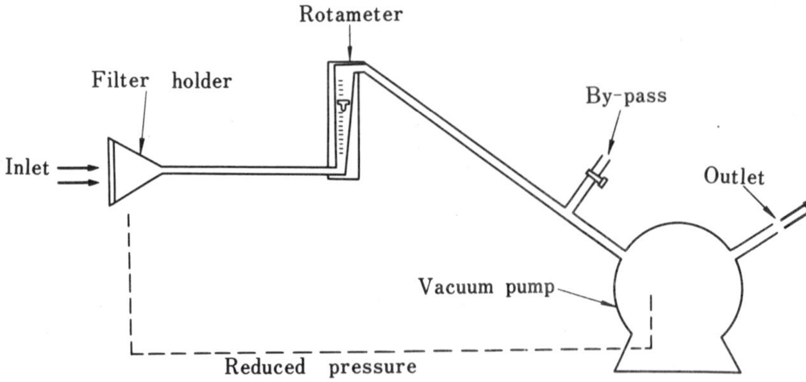

Fig. 2.6. Sketch of the typical structure of a low volume air sampler.

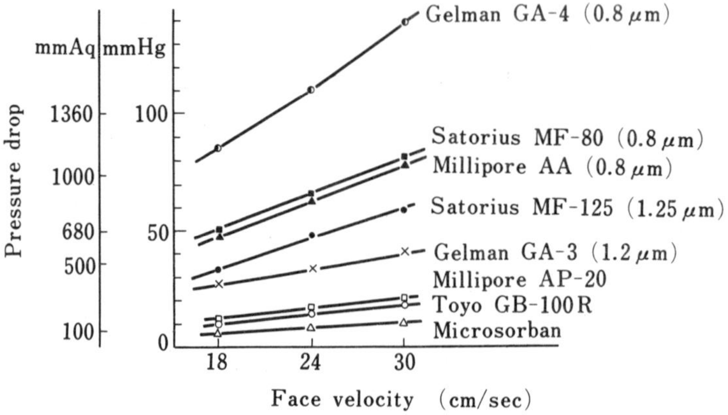

Fig. 2.7. Pressure drop *vs.* face velocity for various filters.

(1) Use of a pressure regulating valve (Fig. 2.8). This is convenient when the flow volume cannot be kept constant due to fluctuations in pump pressure. The valve provides a constant flow volume, regulating the pressure automatically even when the pressure of the flowmeter is changed. Samplers equipped with such valves are available commercially.

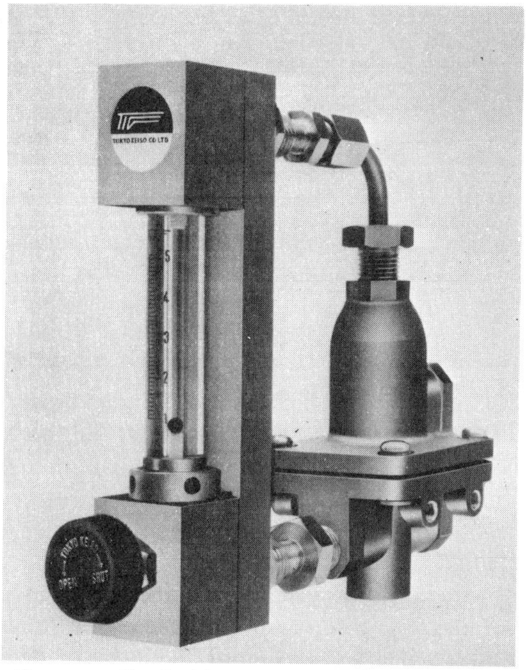

Fig. 2.8. Rotameter fitted with a pressure regulating valve to provide flow control.

(2) Use of a pressure controlling governor. The action of the governor is almost identical to that of the regulating valve just described. Its structure is shown in Fig. 2.9. Gas flows in via a primary port and out via a secondary port passing through a narrow space between the valve and valve sheet. When the pressure at the secondary port is lowered, a negative pressure is transmitted through a pressure guide hole to a diaphragm which is then pulled down, and the valve opens to supply air. If the pressure at the secondary port is elevated, the diaphragm is pushed up to close the valve, gas flow stops and the pressure decreases. Constant pressure is then maintained by repetition of these actions.

(3) Use of equipment which can ensure constant suction volume by increasing motor rotation when the pressure difference measured at an orifice becomes abnormally elevated.

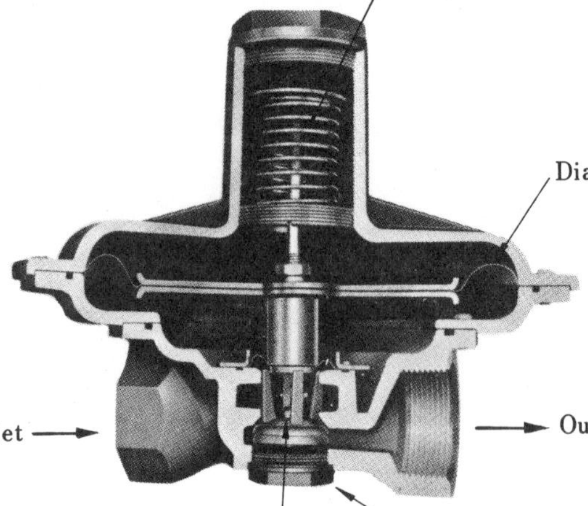

Fig. 2. 9. Sketch of a typical governor.

(4) Electronic monitoring of the pressure difference measured at an orifice, followed by the action of a needle valve in a bypass.

b. Gas meters

There are two types of gas meters, dry and wet test gas meters. The wet test gas meter is used to measure stack gas, while the dry test gas meter is conveniently used when equipped to a high volume or low volume sampler. However, these meters are often misused or misapplied.

i) Wet test gas meters

The principle and basic structure are illustrated in Figs. 2.10 and 2.11, respectively. A U-shaped gas inlet (bent tube in Fig. 2.11) is inserted in a cylindrical drum via a front chamber.

The outer case is filled with water above its mid-point, and the opening of the gas inlet is kept above water level. Gas at higher pressure than that at the outlet enters through the gas inlet, reaches chamber *a* (Fig. 2.10) through a bent tube and so passes into the inner space of the drum. The pressure of the gas in chamber *b* is equal to that at the outlet since it is open to the air. The drum then rotates in the direction indicated by the tailed arrow in Fig. 2.10. In such a way, the gas exhaust is interrupted intermittently as the chambers *a*, *b*, and *c* rotate. When the volume of each chamber is V, gas of volume $3V$ is exhausted per rotation of the drum. The total volme of gas passing through the instrument can thus be estimated by counting the number of drum rotations. The water filling the drum chamber serves both

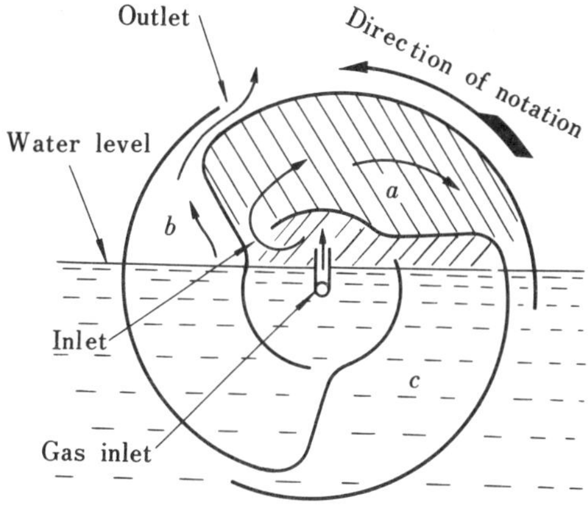

Fig. 2.10. Cross sectional view of a wet test meter.

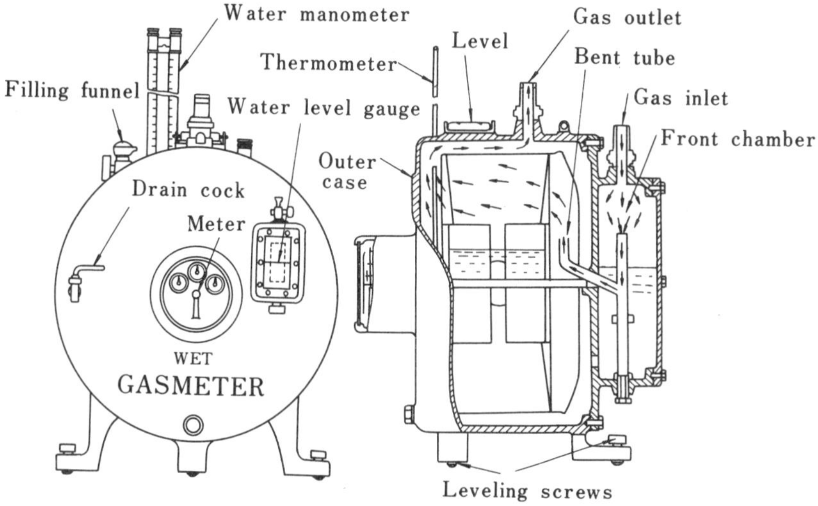

Fig. 2.11. Front and cross sectional views of a wet test meter.

to prevent gas passing through unmeasured, and to push the gas enclosed in each chamber out through its outlet.

The accuracy of the wet test gas meter is so high that it can be referred to as a standard for other flow meters and gas meters. However, errors may result if the water level is not maintained satisfactorily, or the balance between flow volume and pressure reduction is disturbed in either single or double system meters.

One defect of this meter is the increase in weight that accompanies the increase in necessary size when the flow volume to be measured is large.

For example, a meter for measuring a flow volume of as much as 15 l/min may weigh up to 40 kg. This, of course, severely restricts the portability of the instrument. Further, the water must be supplemented or replaced if the meter is transferred, since its level changes and air bubbles are commonly formed.

ii) Dry test gas meters

The structure and principle of action are illustrated in Figs. 2.12 and 2.13, respectively. The main part of the meter is divided into 3 chambers.

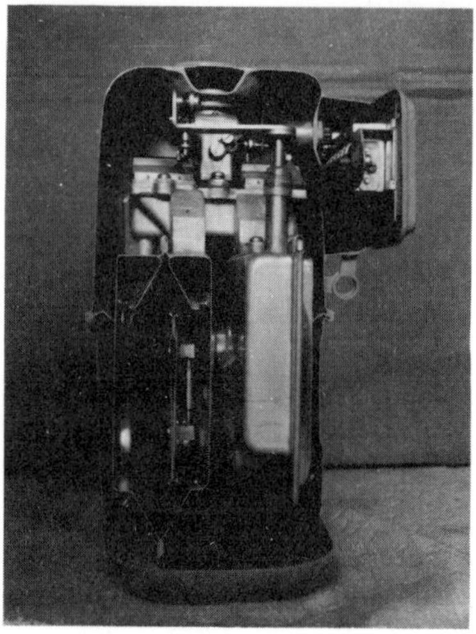

Fig. 2.12. Cross section of a dry gas meter.

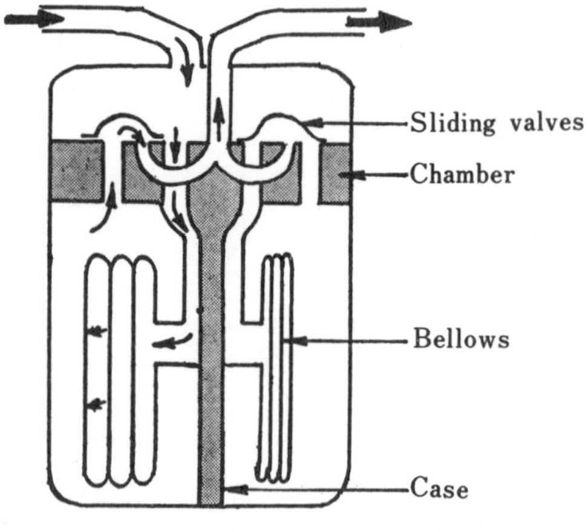

Fig. 2.13. Schematic cross section of a dry gas meter.

The part above the horizontal partition forms a chamber for the sliding valves and integrator, while the lower part is divided into 2 chambers by a vertical partition and contains a pair of bellows. By the action of the sliding valves and valve sheet, gas is supplied to and exhausted from the bellows. The back and forth movements of the bellows are transformed into rotary movements of a crank and the volume is digitally indicated by means of a gear system.

In the case of sampling of suspended particulate matter, the dry test gas meter has been operated under reduced pressure. However, this meter, like the wet test gas meter, should be operated under increased pressure, since especially large errors may result if it is used under reduced pressure. In our experiments, for example, an error of almost 20% was found when the pressure difference was 1000 mm Hg. This defect can easily be realized from the principle of the action as illustrated in Fig. 2.13. As gas is distributed to the bellows by the sliding valves, some gas leakage from the narrow space between the valves and valve sheet is unavoidable. The leaked gas passes directly through the space between the valves and valve sheet to the gas outlet. The volume of leaked gas depends on the size of the space and also on the viscosity of the gas.

C. *Flow meter calibration*

a. Area flow meter calibration

i) Tank method (with gas)

PROCEDURE. Connect a previously calibrated reference tank with the flow meter to be calibrated, as shown in Fig. 2.14. Feed water at a given flow rate from the bottom of the tank for the purpose of removing gas from the top of the tank. Measure the volume of gas removed over a given time or the time taken for removal of a given volume of gas, for comparison with the flow meter indication. Also, measure the gas temperature and pressure in the tank and flow meter for calculation of the gas volume passing through the flow meter as follows.

$$V = \frac{(P_S - P_{SD})T}{(P - P_D)\ T_S} V_S, \qquad (2.2)$$

where V is the gas volume passing through the flow meter over a given time (m^3), V_S the water fed into the tank in a given time (m^3), T_S the absolute temperature of the gas within the tank (°K), T the absolute temperature of the gas in the flow meter (°K), P_S the absolute pressure of the gas within the tank (kg/m^2), P the absolute pressure of the gas in the flow meter (kg/m^2), P_{SD} the saturated aqueous vapor pressure at the temperature T_S

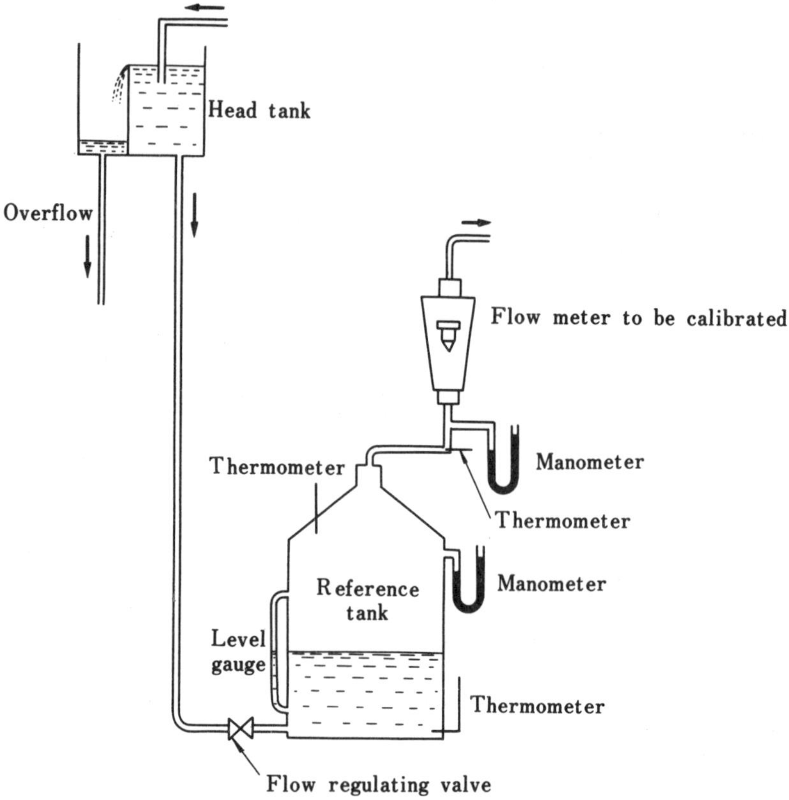

Fig. 2.14. Tank calibration apparatus.

(kg/m^2), and P_D the saturated aqueous vapor pressure at the temperature T (kg/m^2), provided that time, volume, temperature and pressure are measured to an accuracy of within $\pm 1\%$.

Note that the calibrating apparatus should be kept in a place with minimal temperature change and that the temperature difference between the water flowing in the tank and the gas within the tank should be kept as small as possible. Measure the atmospheric pressure if required.

ii) Soap meter method (with a soap film flow meter)

The tank method is disadvantageous since it requires bulky equipment. The soap meter method, which is suitable for measuring gas flow rate especially at pressures near the atmospheric pressure, enjoys a wide range of applications due to the simple instrumental construction and operation. The soap meter consists of a glass cylinder with a uniform inside diameter which is provided with volume graduations (in ml or 1), a rubber tube con-

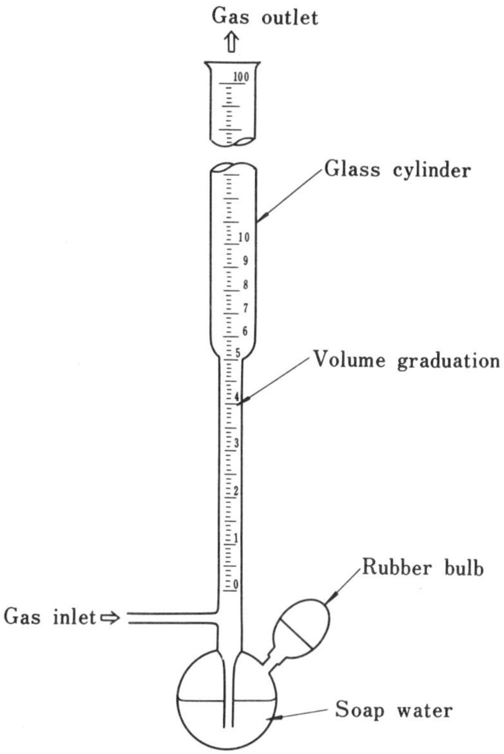

Fig. 2.15. Soap meter (soap film flow meter).

nection at the inlet port and a soap water sump with a rubber bulb at the base, as shown in Fig. 2.15.

PROCEDURE. Gas is introduced under positive or negative pressure via the inlet port near the base and dischanged from the outlet port at the top. Soap film formed by pressing the rubber bulb of the soap water sump is gradually pushed upwards until it breaks at the top of the cylinder. Measure the time taken for the soap film to move from the starting point to an appropriate point on the graduated scale of the cylinder, using a stopwatch, and determine the flow rate in ml/min or l/min. Correct the measured value according to the pressure and temperature at the time.

Soap meters to be used as reference instruments require full instrumental tests.

b. Gas meter calibration

A standard gas meter or gas meter tank (bell-shaped gas holder: spino-

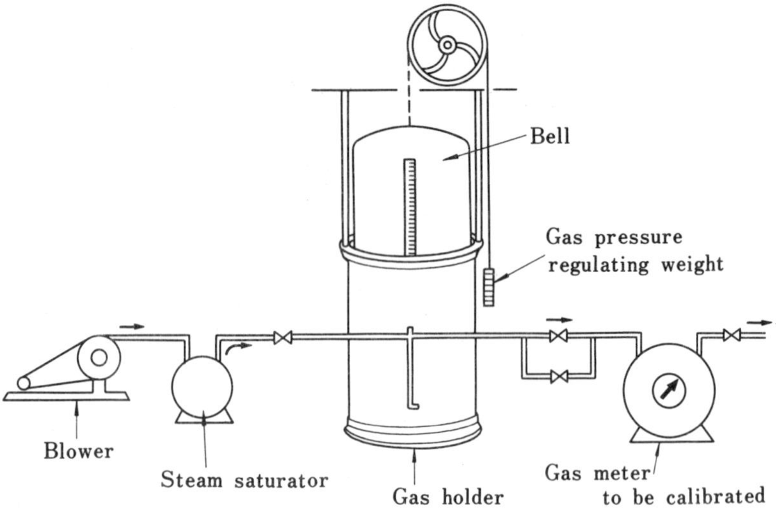

Fig. 2.16. Calibration of gas flow rate with a gas holder.

meter) is generally used for gas meter flow rate calibration. The bell-shaped gas holder shown in Fig. 2.16 consists of a floating bell in a water tank.

PROCEDURE. Charge air under pressure into the gas holder for the purpose of floating the bell at the water level. Then, open the cock to feed gas from the bell to the gas meter to be calibrated. Measure the actual volume of flowing gas and compare it with the flow indication on the gas meter to be calibrated. Then, determine the difference.

It should be noted, however, that the above apparatus is too bulky for an ordinary laboratory, so that gas meters are usually calibrated by connection in series or parallel with a reference gas meter (wet type gas meter in many cases) that has already been calibrated by the above-mentioned procedure.

c. Calibration with a rotary displacement gas meter

The rotary displacement gas meter is a most convenient standard meter for calibrating the flow meter for a suction pump with a capacity as large as 1000–1500 l/min, as used with a high volume air sampler. The general working principle of the rotary displacement gas meter is described here and the calibration procedures are discussed in section 2.2.A.

The rotary displacement gas meter shown in Fig. 2.17 measures the total suction volume with good accuracy. It is a volumetric flow meter that directly measures the flow rate with two rotors. The working principle

Fig. 2.17. A rotary displacement gas meter.

is illustrated in Fig. 2.18. Gas flowing in the direction indicated by the arrow in A produces a pressure difference between the inlet pressure *P* and the outlet in the left rotor and drives it in a clockwise direction. On the other hand, the right rotor is not driven by the pressure difference but is rotated (as shown in B) by the rotating power transmitted by the pilot gears fixed on the shaft until it reaches the position shown in C (a rotation of 90°). In this position, the pressure difference provides the right rotor with rotating power in a clockwise direction. The space (shaded portion) between the rotors and casing serves as the measuring space: one rotation of the rotors discharges gas with 4 times the volume of the measuring space. Thus, the flow rate of the gas passing through the meter can be determined from the volume data transmitted to the integrator.

The basic construction of the rotary displacement gas meter is illustrated in Fig. 2.19. It includes a measuring portion with the two rotors and an indicating portion with the integrator. The volume of the space between the rotors and inner surface of the casing is constant but very small. Usually a meter discharging approximately 0.1 m^3/min of gas at one rotation is used for the calibration of high volume air samplers.

D. Sampling site

The sampler should be installed in a place that constantly represents the average state of the surrounding air pollution and not exposed to the

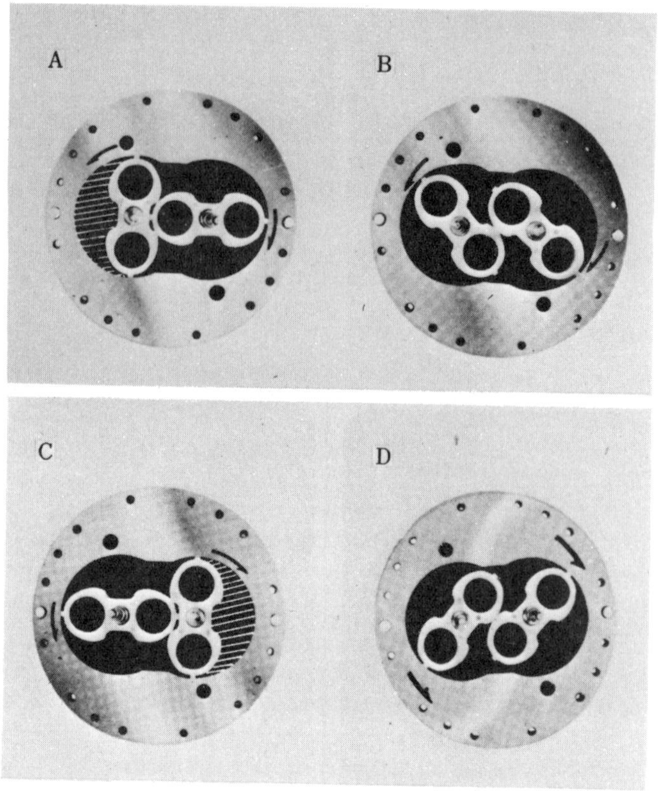

Fig. 2.18. Working principle of the rotary displacement gas meter.

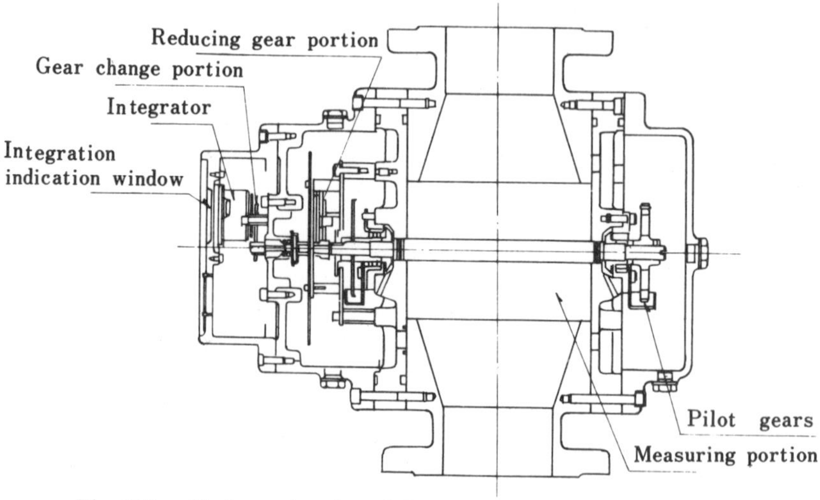

Fig. 2.19. Basic construction of the rotary displacement gas meter.

direct effects of particular discharge sources or traffic. Air samples should be collected at an elevation above the ground of 3–17 m, and preferably at 5–10 m. However, they should be collected from not less than 1.5 m above roofs. Sampling sites paved with pebbles for the prevention of dust generation are recommended.

E. Filter selection

The filter for collecting suspended atmospheric particulates should meet the following requirements: (1) at least 99% collecting efficiency for particulates 0.3 μm and greater in size, (2) low hygroscopicity, since a hygroscopicity exceeding 1 mg/piece leads to serious errors in weight concentration measurement, and so an improper estimate of the environmental concentration, and (3) absence of impurities that might interfere with the analysis.

Glass fiber filters, polystyrene filters and membrane filters are generally used, but cellulose fiber and metal membrane filters are sometimes used for sample collection. The filter material properties, including the physical properties and chemical composition, should be carefully examined. The data in Table 2.2 demonstrate the degree of applicability of various filters generally employed for collecting atmospheric particulates.[1–7)]

a. Glass fiber filters

The filter paper is produced by fixing borosilicate fiber material not thicker than 1 μm with alumina sol $Al_2(SO_4)_3$ or by preparing it in hydrochloric or sulfuric acid solution. Some filters are reinforced with an organic binder such as acrylic resin or Teflon. The softening points depends on the product properties and ranges from 500 to 700°C. The collecting efficiency is measured with 0.3 μm dioctyl phthalate (DOP) smoke. Table 2.3 shows the collecting efficiency and physical properties of various glass fiber filters. Clearly, all such filters exhibit a collecting efficiency of not less than 99.9% for particulates 0.3 μm in size. Fig. 2.20 shows the relation between test aerosol particulate size and the collecting efficiency of various kinds of glass fiber filters, and Fig. 2.21 shows the relation between filtering velocity and the collecting efficiency of Toyo-GB100.

Glass fiber filters contain various metal elements such as Fe, Cr, Co, Ni, Cu, Ti, Cd, Zn, Pb, Rb, Mn, As, Ba, Na, K, etc., besides the main ingredients of Si, B, Al, Ca and Mg (see Table 2.4). As a result, certain metal ingredients significant to health investigations and the identification of pollutant sources cannot be determined in some cases. Gelman Type A and Whatman GF-A contain high levels of impurities which vary considerably from lot to lot. The author has conducted atmospheric investigations around a zinc refinery using a Whatman GF-A filter which had been shown

TABLE 2.2. Properties and applicability of various filters (data from ref. 1–7)

Material	Filter type	Por size (μm)	Electric charge †1	Pressure loss†2	Initial collecting efficiency†3	Hygroscopicity †4	Applicability†4 bor: Low volume air sampler	High volume air sampler	Atomic absorption spectrometry	Activation analysis	Fluorescence X-ray analysis	Remarks
Glass fiber	Toyo GB-100R		○	○	○	○	○	○	○	×	△	
	Gelman type A		○	○	○	○	○	○	○	×	△	High Zn content
	Whatman type A		○	○	○	○	○	○	○	×	△	High Zn content
	Millipore AP-20		○	○	○	○	○	○	○	×	△	High Zn content
	MSA 1106 BH		○	○	○	○	○	○	○	×	△	
Quartz fiber	Tissue Quartz 2,500 QAST		○	○	○	○	○	○	○	△	△	SiO_2 99%
	Gelman Quartz Paper type I		○	○	○	○	○	○	○	△	△	SiO_2 98.5%
	Gelman Quartz Paper type II		○	○	○	○	○	○	○	△	△	SiO_2 99.5%
Polystyrene	Microsorban (sartorius)		○	△	○	△	○	△	○	△	×	
Polyvinyl chloride	Yumicron MF-100	1	○	△	○	○	○	△	○	×	○	High Cl and Ti content
Nitro-cellulose	Millipore-RA	1–3	△	×	○	△	○	×	○	○	○	
	Sartorius SM		△	×	○	△	○	×	○	○	○	
	Toyo TM-100		△	×	○	△	○	×	○	○	○	
Cellulose acetate	Gelman GA-3		△	×	○	△	○	×	○	○	○	
Fluorine resin	Polyfuron PF-3 (Toyo)		○	△	○	○	○	×	△	○	—	Fibrous
	Millipore Mitex LS		○	△	○	○	○	×	△	○	—	Fibrous
	Fluoropore AF-07P (Sumitomo Denko)	10	×	×	○	○	○	△	○	○	—	Membrane type
	Millipore Florinert FA	1	△	△	○	○	○	△	○	○	—	Membrane type

†1 ○, Trace; △, moderate; ×, too high for use.
†2 ○, Very small; △, moderate; ×, large.
†3 ○, Not less than 99%.
†4 Moisture content: ○, 0.1–0.5%; △, 1–2%.
†5 ○, Good; △, sometimes good; ×, not good.

TABLE 2.3. Physical properties and collecting efficiency of various glass fiber filters

filter type / Properties	Gelman type A	Whatman GF–A	Whatman GF–C	MSA 1106 BH	Toyo GB–100R
Thickness (mm)	0.27	0.19	0.19	0.24	0.33
Weight (g/m²)	82.2	66	60	66	93.9
Air permeability (sec/cm)	22.4	20.8	31.8	19.7	18.7
Tensile strength (kg/15 mm)	0.35	0.09	0.25	0.14	0.12
Pore size (μm)	13.3	8.2	7.2	12.2	8
Collecting efficiency for 0.3 μm DOP smoke (%)	99.9	99.9	99.9	99.99	99.9

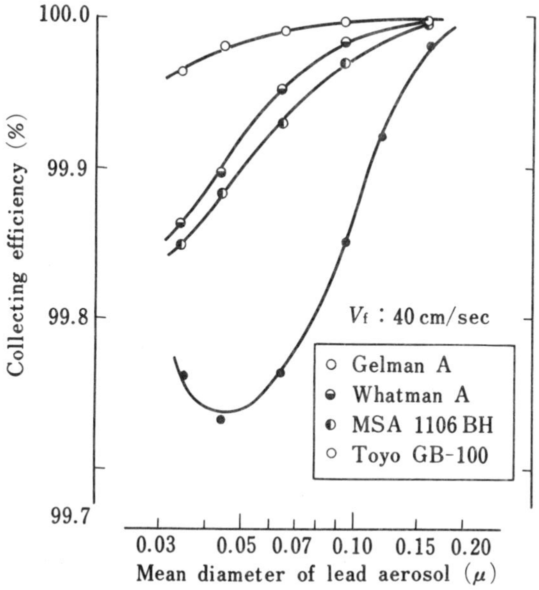

Fig. 2.20. Relation between glass fiber filter paper collecting efficiency and test aerosol particulate size.[4]

to contain small amounts of impurity by a blank test. The analysis revealed far more zinc than expected, indicating that the zinc measurements were inaccurate. Later tests clarified that a filter containing more impurities (including zinc) than any of the other products had been chosen for the measurement from a lot other than that (on the same delivery) selected for the blank test. Since products from the same maker may thus differ in quality, it is important to choose a filter from the same lot as that subjected to the blank test prior to commencing investigations.

Toyo GB-100R is produced from Japan Glass Fiber 4A (the material used for the conventional Toyo GB-100) by the addition of 5% of AAA-

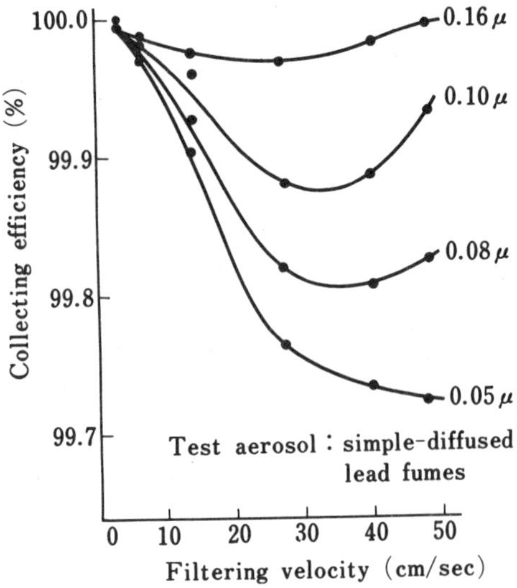

Fig. 2.21. Relation between filtering velocity and collecting efficiency of Toyo GB–100.[4)]

10 from Owens Corning Co., U.S.A. The filter is purified by dipping in 3 N HCL for 24 h and washing with purified water until neutral. The filter has not less than 99.9% collecting efficiency for 0.3 μm DOP smoke and is the lowest in metal impurity content, especially zinc content.

Careful attention should be paid to the recent finding that glass fiber filters adsorb acidic gases such as sulfurous acid gas well, leading to a considerable positive error in particulate concentration measurements. As mentioned, glass fiber filters contain not only certain heavy metals to be measured but also alkali metals including sodium, potassium, etc. The latter may react, for example, with sulfurous acid gas during atmospheric sampling to produce sodium sulfate, etc., resulting in an apparent, spurious weight gain in particulate amount ranging from 20 to 40% depending on the kind of filter used. Hence, glass fiber filters should in general not be used for metal or weight concentration measurements.

On the other hand, quartz fiber filters do not adsorb acidic gases since they are composed of 99% silicate and contain no alkali or alkali earth metals, although they do contain minimal amounts of metal impurities as shown in Table 2.4. For these reasons, quartz fiber filters will probably replace glass fiber filters in the future.

b. Polystyrene filters

Polystyrene filters are applicable to high volume air samplers having a 110 mm suction diameter and 600–700 1/min capacity since their resistance

TABLE 2.4. Metal content of glass fiber and quartz fiber filters as determined by atomic absorption spectrometry

Filter type	GB-100R		MSA 1106 BH		Gelman type A		Whatman GF-A†1		Whatman GF-A†2		Palleflex 2500 GAS	
Material	glass fiber		glass fiber		glass fiber		glass fiber		glass fiber		quartz fiber	
Maker	Toyo Roshi Co.		Mine Safety Appliance Co.		Gelman Instrument Co.		W. & R. Balston		W. & R. Balston		Palleflex	
Content / Element	μg/14 in²	μg/cm²	μg/14 in²	μg/cm²	μg/14 in²	μg/cm²	μg/14 in²	μg/cm²	μg/14 in²	μg/cm²	μg/14 in²	μg/cm²
Fe	55.2	0.611	85	0.94	64	0.71	220	2.43	120	1.33	10	0.11
Ni	4	0.04	7	0.08	1	0.01	7.5	0.083	5.2	0.057	<0.7	<0.008
Mn	2.1	0.023	2.5	0.028	1.5	0.017	15	0.17	10	0.11	1.45	0.016
Cr	2.2	0.024	3.2	0.035	1.8	0.020	5	0.05	3.3	0.036	<0.1†2	<0.001
Sb	—	—	20	0.22	15	0.17	30	0.33	17	0.19	—	
Pb	4.5	0.05	50	0.55	10	0.11	40	0.44	15	0.17	1.9	0.021
Zn	9	0.1	15	0.17	4625	51.20	25000	276.8	3175	35.15	1.2	0.013
Cd	<0.1	<0.001	<0.1	<0.001	<0.1	<0.001	<0.1	<0.001	<0.1	<0.0001	<0.18	0.0020
Cu	1.5	0.017	7	0.08	1.1	0.012	2.5	0.027	1.8	0.020	1.2	0.013
Ca	280	3.10	900	9.96	550	6.09	3250	35.98	1250	13.84	—	
Mg	252	2.79	650	7.19	320	3.54	800	8.85	300	3.32	—	
Na	5600	62.0	3000	33.21	3150	34.87	20500	226.9	10000	110.7	—	
K	395	4.37	275	3.04	325	3.60	415	4.59	840	9.30	—	

†1 Received in October, 1969.
†2 Received in October, 1970.
†3 0.35 μg/14 in² (0.0036 μg/cm²) on alkali fusion treatment.

to air permeation is small. The Microsorban Filter of Sartorius Membrane Filter Co. is the most frequently used commerical product due to its low metal impurity content. However, particular attention is necessary during weighing, since this filter is highly hygroscopic.

c. Membrane filters

Membrane filters consist of cellulose derivatives, usually nitro-cellulose or acetyl-cellulose. Products with various pore sizes are available but the individual pore size is quite uniform, so that a particular size can be selected according to the usage. A pore size of 0.8 μm is suitable for suspended dust collection.

The ash content is generally low, i.e. 10^{-5} to 10^{-4} mg/piece of 47 mm diameter filter with a 0.8 μm pore size. However, some filters with a high ash content, and variability in ash content between different lots are found. For this reason, the metal content, which differs by maker and lot

TABLE 2.5 Background metal content of membrane filters for neutron activation analysis (units, ng/cm^2)

Filter type / Element	Microsorban (Sartorius)[8]	Toyo TM-300[9]	Millipore AA[7]	Gelman GA[7]	Sartorius SM[10]
C	260	190	1,000	600	240
Br	ND	1.0	2	4	5.8
Na	7.6	500	400	2,200	130
K	ND	36	100	—	—
Mg	ND	70	200	—	360
Ca	41	300	370		1,250
Al	11	260	10	740	80
Sc	ND	0.02	<0.05	—	0.005
Ti	ND	—	< 10	—	3.1
Fe	ND	30	40	—	40
Mn	0.095	1.2	2	2	1
Co	ND	0.2	0.1	—	0.2
Ni	—	—	< 20	—	1
Ag	ND	0.5	< 1	—	0.2
Cu	4	8	60	30	—
Zn	730	6	7	—	5
Sb	0.20	0.04	1	—	0.45
Cr	ND	0.7	5	—	1.7
Hg	1.0	—	0.5	—	0.2
V	ND	—	<0.05	0.05	0.7

(see Table 2.5), should be examined before use, especially in the case of neutron activation analysis.

Membrane filters are highly resistant to water, weak alkali and weak acid, but soluble in ketone, ester, ether and alcohol. Taking advantage of such solubility, the metal content of suspended dust can be determined by dissolving the membrane filter carrying the collected dust in acetone, evaporating the acetone, and then dissolving in acid. These filters are not so resistant to heat (limit, about 125°C).

Direct insertion of the membrane filter into the muffle furnace for ashing results in explosive combustion, causing ash loss. To avoid this, the membrane filters should be dipped in isopropyl alcohol prior to ashing.

d. Metal membrane filters

As mentioned above, glass fiber filters contain various metal contents in addition to their main ingredient, borosilicate, and sometimes the amounts are so great that a certain metal element or elements cannot be determined. Membrane filters are liable to break unless special care is taken during handling, and cellulose membrane filters sometimes contain abundant impurities, especially metals. On the other hand, the silver membrane filter with 99.99% purity which has recently been developed in the U.S.A., contains minimal amounts of impurities and can be used in the same way as the cellulose membrane filters. Products with various pore sizes are available from Selas Flotronics, Co., but those with a 1.2 or 0.8 μm pore size are generally used. 100% recovery has been attained in desorbing dust collected on the filter of a low volume air sampler, by application of approximately 30 ppm nonionic surfactant and using an approximately 200 Hz supersonic wave. The desorbed dust was dissolved in 20 ml of nitric acid (1:1) and 5 ml of 30% hydrogen peroxide in a flask with a reflex condenser. This filter is most suitable for collecting atmospheric lead.

e. Cellulose fiber filters

Cellulose fiber filters have not been used widely due to their high hygroscopicity and low collecting efficiency. However, their applicability has recently been reviewed due to their negligible metal impurity content and the satisfactory particulate collecting efficiency that can be attained by adjusting the filtering velocity. They are especially useful in the case of sampling for neutron activation analysis.

Fig. 2.22 shows the relation between filtering velocity and collecting efficiency of Whatman No. 41 filter for polystyrene particulates of 0.264 μm diameter. Clearly, the Whatman No. 41 filter (47 mm in diameter) attains a collecting efficiency of approximately 90% in the case of a low volume air sampler and approximately 95% in the case of a high volume air sampler.

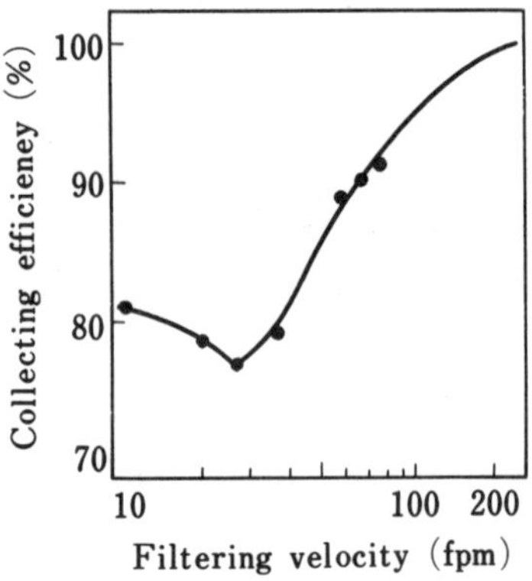

Fig. 2.22. Relation between filtering velocity and collecting efficiency of Whatman No. 41 filter for 0.264 μm polystyrene particulates.
(Source: ref. 12. Reproduced by kind permission of Pergamon Press Ltd., England.)

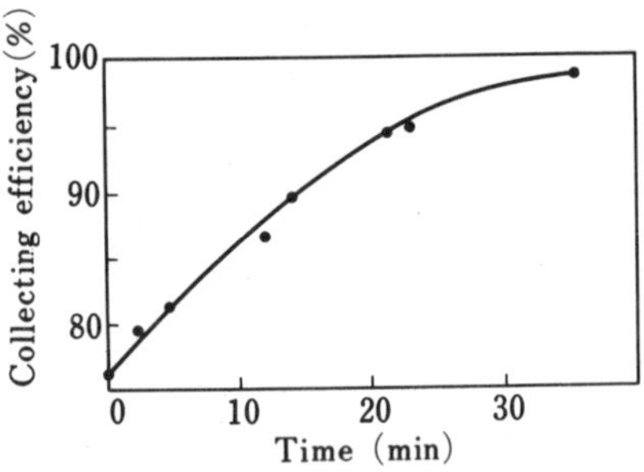

Fig. 2.23. Collecting efficiency change of Whatman No. 41 filter for 0.365 μm particulates at a density in air of 0.5 mg/m^3, resulting from clogging in use at a filtering velocity of 30 fpm.
(Source: ref. 12. Reproduced by kind permission of Pergamon Press Ltd., England.)

Since the filter is clogged during application, its collecting efficiency in an experiment conducted with 0.365 μm polystyrene particulate at a density of 0.5 mg/m^3 in air, increased from 75% at first to not less than 93% after 30 min even at a filtering velocity of 30 fpm, which gave the lowest collecting efficiency, as shown in Fig. 2.23.

Cellulose fiber filters are nevertheless too hygroscopic for accurate weighing and the precision of weight concentration measurement is severely reduced. Table 2.6 lists the hygroscopicity of various kinds of filters. On the other hand, since the metal content of cellulose fiber filters is mostly low, as shown in Table 2.5, and an appropriate filtering velocity giving

TABLE 2.6 Properties of various kinds of sampling filters[4)]

Material	Weight† (mg)	Thickness (mm)	Packing density (%)	Hygroscopicity (%)
Cellulose	172–216	0.24–0.29	23–30	8.4–11.1
Cellulose-asbestos	164–250	0.42–0.50	13–17	3.0–8.3
Glass fiber	135–299	0.36–9.97	6–8	0.1–0.4
Membrane	60–103	0.053–0.091	18–40	1.3–1.8

†Circular piece of filter 55 mm in diameter.

maximum collecting efficiency can be selected, they the represent an excellent kind of filtering medium in general.

F. Weighing

PROCEDURE. Weigh the glass fiber filter after allowing it to stand for 16–24 h in a weighing room maintained at a temperature of 24°C and humidity of 50%. Collect the dust sample and allow the filter to stand under the same conditions as above until constant weight is reached. (Since the filter does not attain constant weight as readily as before, it is recommended that it be allowed to stand for 48 h in this case.) Then, weigh the filter again. Note also that it is important to reserve 10% of the pieces weighed before collecting the dust, for the purpose of confirming the same weight as initially, when they are weighed simultaneously with the pieces serving for dust collection.

Since it is difficult to maintain the same conditions for a long time even in a room kept at constant temperature and humidity, an accurate weight concentration is difficult to achieve.

In the case where the samples are few or no thermo-hygrostatic room at 24°C and 50% humidity is available, the following weighing produre is recommended. Place two pieces of clean glass fiber filter in a desiccator with magnesium perchlorate desiccant and allow the desiccator to stand in a weighing room. Take out one of the pieces of filter from the desiccator and quickly weigh it while measuring the time with a stopwatch. Record the weight change with time and draw a weight change curve. Repeat the same procedure with the second piece. Derive the weight of each piece immediately after removal from the desiccator, by extrapolating each of the curves to the ordinate axis, as shown in Fig. 2.24. Repeat the drying and weighing procedure at an interval of 3 h until the initial weight reaches the same value. The procedure giving the most accurate value is that adopted in the British Standards (B.S. 1747–2).

The commonly marketed direct indication balance is inconvenient for weighing 8 × 10 in glass fiber filters since the small, circular weighing plate does not provide a stable base on which to nest the filter. If a balance with-

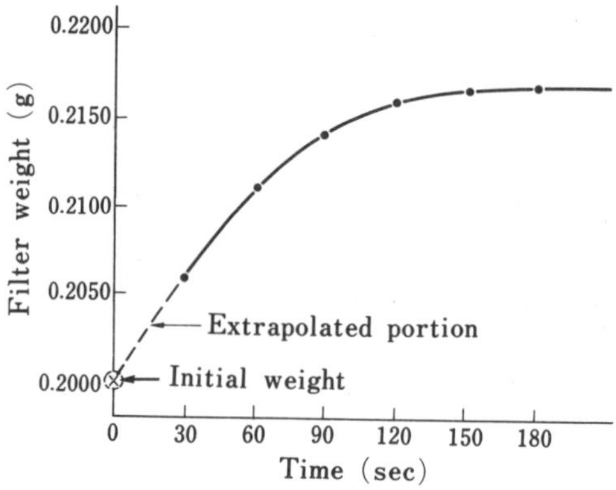

Fig. 2.24. Determination of filter weight by extrapolation.

out a weighing chamber capable of containing filters of this size is used, the filter may be folded or bent, but this occasionally causes damage. A balance provided with accessory units on both sides (see Fig. 2.25), which is sufficiently large to contain an 8 × 10 in filter, is now available on the market.

The method of weighing a sample after drying at 100°C and allowing

Fig. 2.25. Balance provided with accessory units on both sides for accomodating an 8 × 10 in filter.

it to stand in a desiccator is frequently used; however, it is not applicable to measurements of suspended dust concentrations. The sample (on a collecting filter) dried in a silica gel desiccator for 48 h measures 10% heavier than that allowed to stand for 40 min immediately after drying in a drier (at 100°C) for 2 h. The reason is that a portion of the organics in the collected sample is evaporated by drying at 100°C. Hence, allowing the sample to stand in a silica gel desiccator for not less than 48 h, rather than drying at 100°C, is recommended for determining the gross weight of suspended dust.

2.2. Atmospheric particulate sampling

The sampling method employed depends on whether the sampled suspended dust is to serve for chemical analysis, weighing or counting. The methods related to chemical analysis are summarized here.

(1) Impinger method. The sample is collected by sucking air at a rate of 1–30 l/min through a nozzle using a suction pump and bubbling through water or absorbent solution contained in a glass cylinder. Atmospheric lead is frequently collected at a sucking rate of 30 l/min in a 100 ml impinger containing 75 ml of 1% nitric acid.

(2) Filter method. As described in the precious section, the sample is collected with a high volume sampler (using a glass fiber or polystyrene filter), with a low volume sampler (using a membrane or other suitable filtcr), or with certain other kinds of samplers.

(3) Electric dust collector method. This method utilizes an electrostatic dust sampler. The samper consists of a sampler body, flow meter, suction pump and high voltage power source, as shown in Fig. 2.26. A high voltage is applied between the discharging electrode and duct-collecting electrode

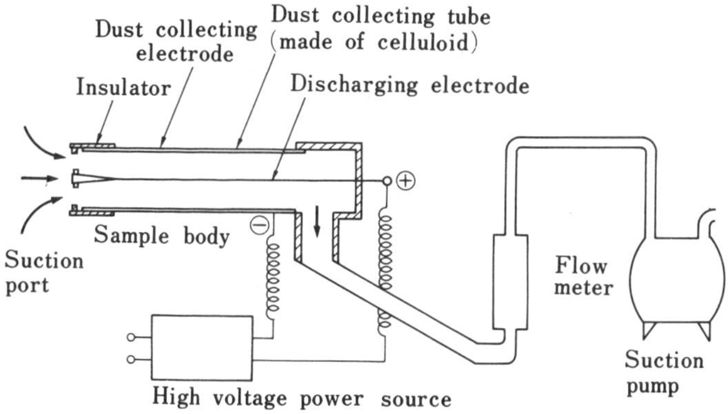

Fig. 2.26. Electrostatic dust sampler.

to cause a corona discharge within the dust-collecting tube. The charged dust with ions generated in the tube is attracted onto the inner wall and collected.

(4) The Andersen sampler and cascade centripetal sampler designed for determining particle size distributions but usable for chemical analysis, exist in addition to the above-mentioned samplers.

In this section, collecting methods using high volume, low volume and Andersen samplers will be considered mainly.

A. *High volume air sampler*

a. General

In the method of collecting atmospheric suspended particulate matter on a filter with a high volume air sampler, particulate matter is generally collected by sucking air at a flow rate of approximately 1.2–1.7 m^3/min. This method provides a sampler for the analysis of metal and other ingredients.

b. Apparatus

The high volume air sampler consists of an air suction assembly, filter holder, flow rate measuring assembly and shelter (see Figs. 2.27 and 2. 28).

i) Air suction assembly

The air suction assembly consists of a series motor and dual centrifugal turbine fan which are directly connected. It sucks air at a rate of approximately 2 m^3/min under no load and is capable of continuous operation for not less than 24 h.

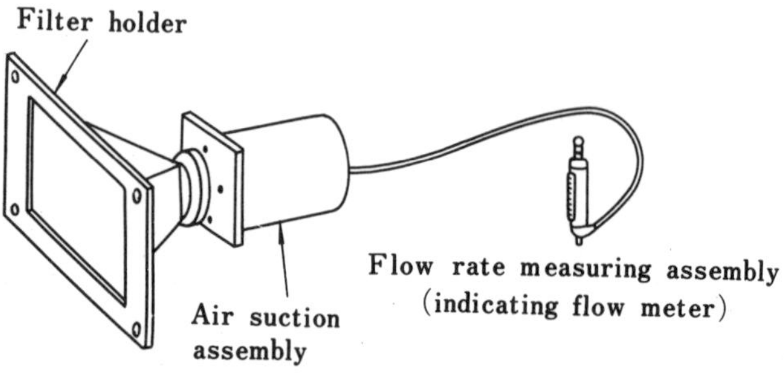

Fig. 2.27. High volume air sampler.

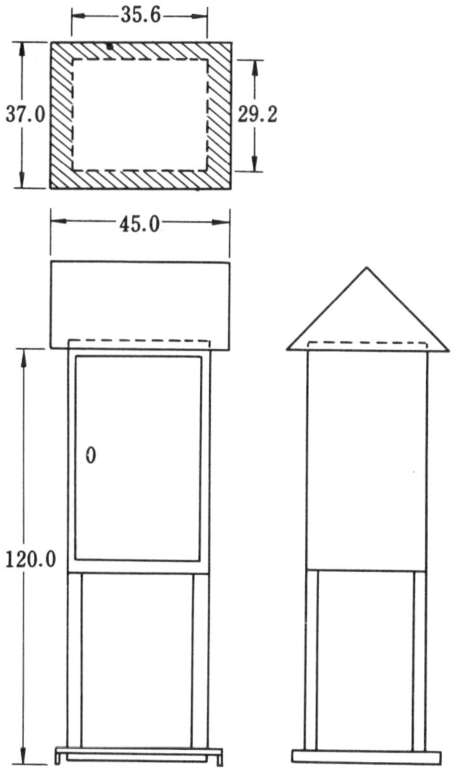

Fig. 2.28. Example of a shelter for a high volume air sampler.

ii) Filter holder (Fig. 2.29)

A piece of filter approximately 20 × 25 cm^2 (8 × 10 in^2) in size can generally be fitted into the filter holder without break or air leak, and the filter holder is directly connected to the air suction assembly. The material and dimensions of the components of the filter holder are as follows.

(1) Frame: made of acid-resistant material, with outer dimensions of 24 cm × 29 cm and inner dimensions of 18 cm × 23 cm, which provides a fixture for the specified-size filter without causing breaking.

(2) Net: made of anticorrosive material which supplies no contaminants to the filter, is strong enough to support the filter without breaking during the feeding of air and has the same dimensions as the filter. Fluorine resin tape is applied to the portion of the frame where the air flow is to be screened.

(2) Packing: made of synthetic independent-foam rubber and having

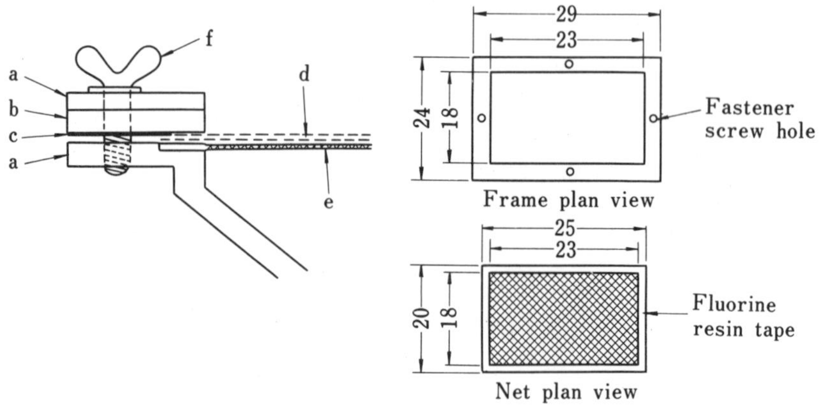

Fig. 2.29. Filter holder assembly. a, Frame: b, packing; c, fluorine resin tape: d, filter; e, net; f, fastener.

the same dimensions as the frame. Fluorine resin tape is applied to the portion in contact with the filter.

(4) Fastener: made of anticorrosive material, and designed to prevent air leaks and filter breaks after the filter has been fitted.

iii) Flow rate measuring assembly

The flow rate is generally measured with an indicating flow meter, i.e. a float flow meter combined with the air suction assembly and ready to be engaged or disengaged quickly. The flow meter has relative flow rate unit graduations covering the range from 1.0 to 2.0 m^3/min to an accuracy of 0.05 m^3/min. The graduations should be calibrated with a standard flow meter under normal operating conditions of the high volume air sampler.

iv) Shelter

The shelter is fixed level, with the dust collecting surface directed upwards. It serves to protect the filter from the wind or rain and is made of anticorrosive material. It consists of a roof, case body and legs, and the area of the space between the roof edge and case body (shaded portion in Fig. 2.28) measures 650 ± 65 cm^2, while that of the rectangular portion at the case body top is approximately 30×35 cm^2.

c. Procedure for collecting particulate matter

i) Filter weighing before collection

Accurately weigh the filter to the nearest 0.1 mg with a chemical balance having a 0.01 mg sensitivity, after the filter has reached constant weight at a temperature of 20°C and relative humidity of 50%.

ii) Collection of particulate matter

(1) Check the sampler to confirm that it operates normally.

(2) Fix the weighed filter to the filter holder so that all air leaks are prevented.

(3) Fix the sampler level, with the filter surface directed upwards within the shelter.

(4) Connect the flow meter to the tap provided on the discharge panel of the sampler back, using a rubber tube.

(5) Switch the power source on and record the starting time of collection.

(6) After 5 min of collection, read and record the flow meter at the float center and then disengage the flow meter from the sampler.

(7) At the scheduled time of completion of collection, again connect the flow meter to the sampler and record the flow meter reading as before.

Next, calculate the air suction (m^3) from the following equation, based on the flow meter readings at the start and end of collection.

$$\text{Air suction} = \frac{Q_s + Q_e}{2} T, \tag{2.3}$$

where Q_s is the flow rate at the start of collection (m^3/min), Q_e the flow rate at the end of collection (m^3/min), and T the collecting time (min).

iii) Weighing of filter after collection

Fold the filter parallel to the shorter sides (approximately 20 cm in length) with the particulate collecting surface inside, allow the filter to stand for at least 24 h at a constant temperature of 20°C and relative humiditly of 50%, and then weigh.

iv) Data to be recorded

Record the sampling place and date, filter number, time at the start and end of sampling, useful meteorological data (weather condition, temperature, humidity, wind direction, wind velocity, etc.) and collector's name for each collected sample.

v) Collecting filter

A filter with a collecting efficiency of not less than 99% for 0.3 μm aerosol, low pressure loss and hygroscopicity, and no content of substances interferring with the analysis should be used for collecting the particulate matter. Glass fiber filters having the above-mentioned physical and chemical properties are generally used. However, since all such filters do not meet all the above requirements, care is required in filter selection (*cf.* Tables 2.2–2.4).

Note: The filtering velocity of an ordinary high volume air sampler is 0.5 m/sec (1.5 $m^3/min/500\ cm^2$). The efficiency for collecting 0.3 μm aerosol particulates by inertial collision or diffuse aggregation is said to be lowest at around the filtering velocity. Since a filter with a collecting efficiency of not less than 99% for 0.3 μm aerosol particulates collects not less than 99% of the particulates smaller than 0.3 μm (with a greater diffuse aggregation effect) or larger than 0.3 μm (with a greater inertial collision effect), a particulate size of 0.3 μm is specified for the test aerosol when estimating the collecting efficiency.[14)]

d. Calculation of the particulate matter concentration

The particulate matter concentration is calculated from the filter weight as determined in c. i) and iii) and the air suction as determined in c. ii), using the following equation.

$$\text{Particulate matter concentration} = \frac{W_e - W_S}{V} \times 10, \qquad (2.4)$$

where W_e is the filter weight after sampling (mg), W_S the filter weight before sampling (mg), and V the air suction (m^3).

e. Filter division

The portion of the filter required for analysis is cut from a part not including the central fold line. The remainder of the filter is preserved in a cool, dark place free from moisture

f. Flow rate correction

Correction of the accessory flow meter reading to the true flow rate is necessary when sampling dust with a high volume air sampler The orifice calibrated using a rotary displacement gas meter as the standard flow meter is used for this correction.

i) Apparatus (Fig. 2.30)

The apparatus used consists of a rotary displacement gas meter, flow rate correcting orifice, manometer 5 types of air diffusers (A–E, metal disks 0.24 cm in thickness and 9.2 cm in diameter, with 1–9 holes 12 mn in diameter, of increasing resistance from A to E), the sampler body, a thermometer and barometer.

ii) Orifice calibration

(1) Assembly. The orifice with the manometer is fitted to the air inlet port of the rotary displacement gas meter and the high volume air sampler body is connected to the Roots meter outlet port with a connector into which the air diffusers can be inserted.

(2) Procedure

(a) Switch on the power source, operate the apparatus for 5 min under no load (idling), and record the rotary displacement gas meter indication and manometer reading during this 5 min of operation.

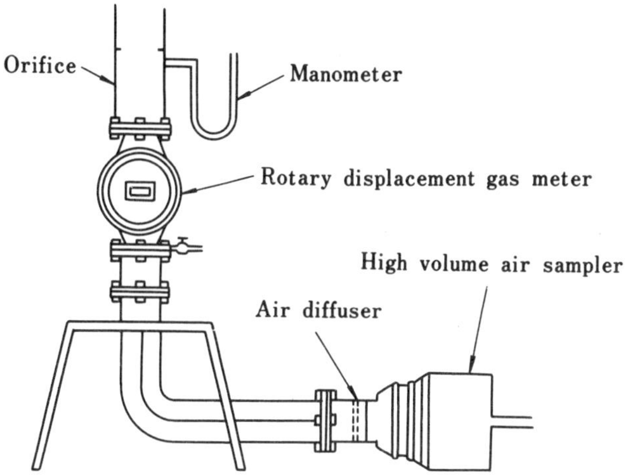

Fig. 2.30. Orifice evaluating assembly.

(b) Turn the power source off and insert the air diffuser of lowest resistance (A) between the rotary displacement gas meter and high volume air sampler body. Put the power on again and record the rotary displacement gas meter indication and manometer reading for 5 min of operation as in (a). Then, repeat the same procedure for the air diffusers B to E, and also measure the atmospheric temperature and pressure.

(c) Correct the recorded rotary displacement gas meter indication according to the measured atmospheric temperature and pressure (20°C, 1 atm) and prepare a calibration curve from the corrected indication and manometer readings (see Fig. 2.31), using the correcting equation,

$$Q' = Q \times \frac{293}{273 + t} \times \frac{P}{760}, \qquad (2.5)$$

where Q' is the true flow rate (m^3/min), Q the flow rate indication on the rotary displacement gas meter (m^3/min), t the atmospheric temperature (°C), and P the atmospheric pressure (millibar).

iii) Accessory flow meter calibration

(1) Assembly (see Fig. 2.32)

The calibrated orifice with the manometer is fitted to inlet port of the high volume air sampler.

(2) Procedure

a) Switch the power source on and operate the apparatus for 5 min under no load to read the manometer and accessory flow meter indications.

b) Switch the power source off and insert air diffuser A into the connecting portion between the orifice and high volume air sampler body.

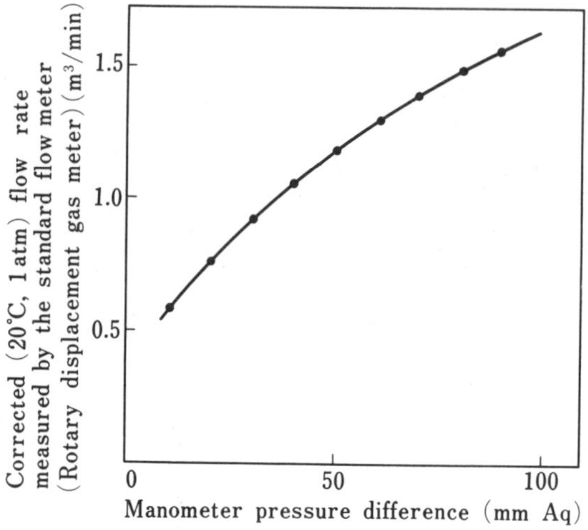

Fig. 2.31. Orifice calibration curve.

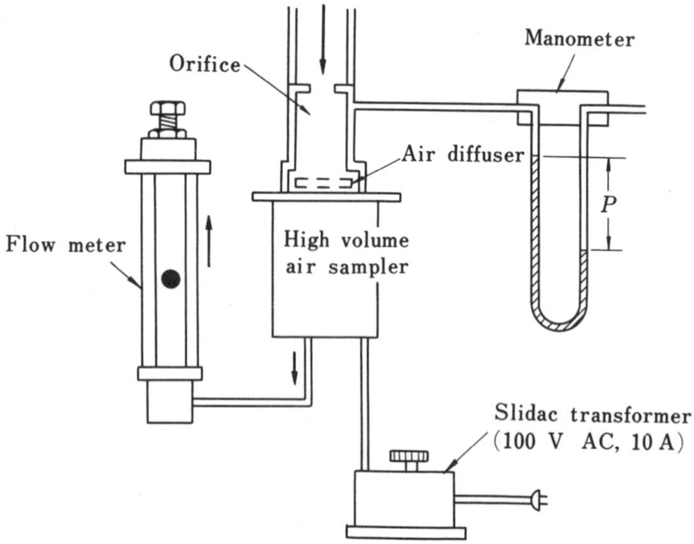

Fig. 2.32. Flow meter calibration assembly.

Again the switch power source on and operate the apparatus as in (a) for 5 min to read the manometer and accessory flow meter indications. Repeat the above-mentioned procedures for air diffusers B to E.

c) Derive the true flow rate from the manometer indications on air diffuser insertion and the calibration curve prepared above (Fig. 2. 32).

d) Plot the accessory flow meter indication against the true flow rate derived in (c) to obtain a graph (Fig. 2.33). Correct the flow rate indicated on the accessory flow meter, using the graph.

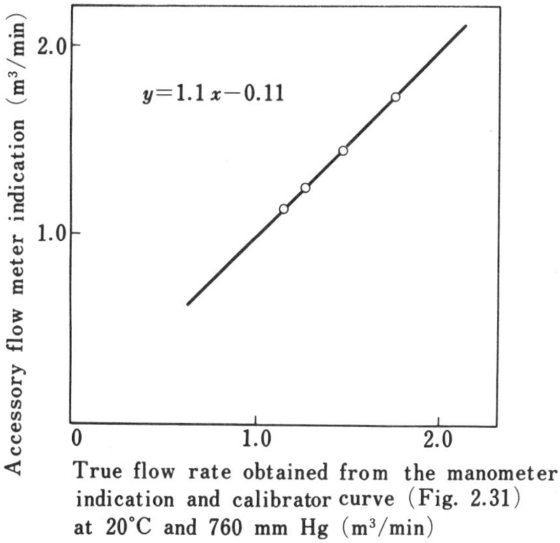

Fig. 2.33. Relation between accessory flow meter indication and true flow rate.

g. Notes

(1) If the flow rate measurement during sample collection or weight concentration measurement after sample collection gives an abnormal value, the flow meter should be inspected for operational irregularities, the sampler for air leaks and the power source for voltage fluctuations. If the abnormality occurs immediately after commencing sampling, sampling should be re-started after confirming recovery to normal operation. If the abnormality occurs at the end of sampling, sampling should be re-started under sufficient control for the prevention of any reappearance of the abnormality, and the entire sample and detailed records kept.

(2) The motor brush of the suction unit should be replaced after a serving time of 400–500 h (or 17–20 runs of 24 h continuous duration) and flow rate calibration be conducted upon replacement.

(3) The flow rate regulating screw at the upper end of the accessory flow meter should be fixed, since even slight deviations from the fixed position require flow meter calibration with the orifice.

(4) Dirt adhering to the narrow portion at the upper end of the accessory flow meter leads to a smaller reading. Such dirt should therefore be removed with a fine wire without scratching. The flow meter should be calibrated with the orifice after each removal of dirt.

(5) The flow meter should also be calibrated with the orifice on replacement or repair of suction unit components or abnormal flow rate indications during sampling.

B. Low volume air sampler

a. General

This section describes the methodology for collecting atmospheric particulate matter on a filter usually at a suction rate of 20–30 1/min for metal content analysis.

b. Apparatus

The low volume air sampler consists basically of a pump, filter holder and flow rate measuring assembly (see Fig. 2.34).

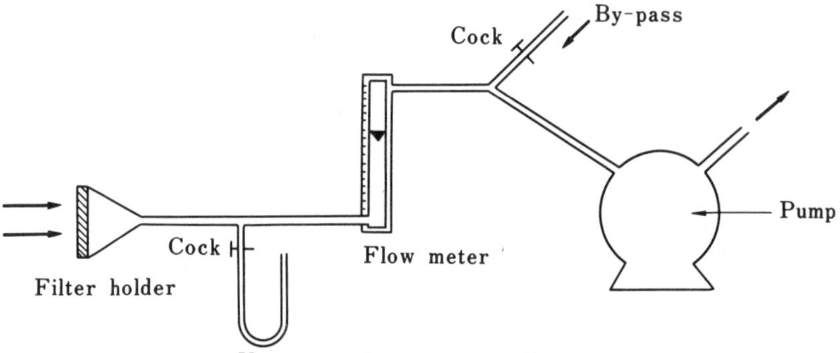

Fig. 2.34. Construction of the low volume air sampler.

i) Suction pump

An eccentric rotor pump or diaphragm pump durable to not less than 30-day continuous operation is used.

Note: the requirements for the suction unit are (1) attainment of a high vacuum, (2) a large suction flow rate, (3) no pulsations, and (4) portable construction.

ii) Filter holder

A filter holder which fits the filter without break or air leak, is used (see Fig. 2.35). Its components are as follows.

(1) Frame: made of anticorrosive material.

(2) Net: made of anticorrosive material which supplies no contaminants to the filter, and is strong enough to support the filter without breaking during the feeding of air.

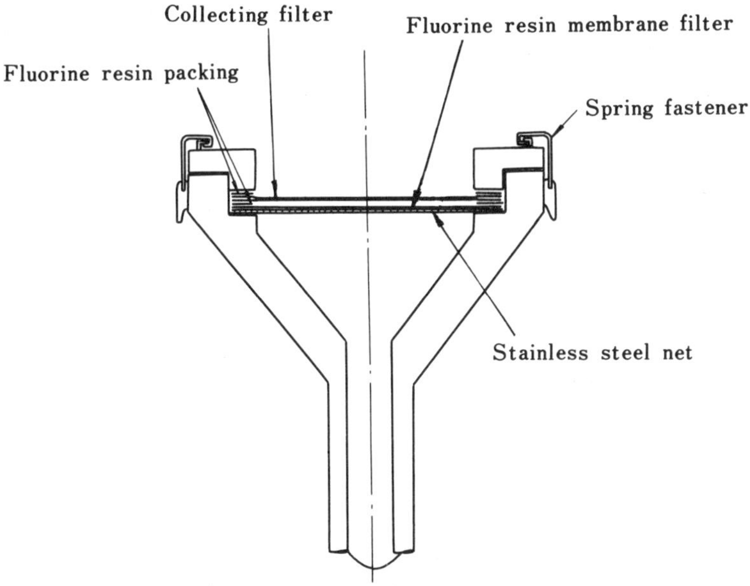

Fig. 2.35. Filter holder assembly.

(3) Packing: made of fluorine resin.

(4) Fastener: made of anticorrosive material, and designed to prevent air leaks and filter breaks.

iii) Flow rate measuring assembly

The flow rate is measured with a float area flow meter having a range of from 10 to 30 l/min, and accuracy of 0.5 ml/min at 20°C and 1 atm, which is set between the filter holder and suction pump.

iv) Collecting filter

A filter with an initial collecting efficiency of not less than 99%, low pressure loss, hygroscopicity and electric charge, sufficient handling strength, and no content of substances interfering with the analysis should be used for collecting the particulate matter. Glass fiber filters or nitrocellulose or acetyl-cellulose membrane filters with a 1–3 μm pore size are generally used. However, glass fiber filters composed mainly of ingredients interfering with the analysis of sodium, etc. are not suitable in the case of sampling for neutron activation analysis. Care is thus required in filter selection (*cf.* Tables 2.2–2.4).

c. Procedure for collecting suspended particulate matter

i) Filter weighing before collection

Accurately weigh the filter to the nearest 0.1 mg with a chemical balance having a 0.01 mg sensitivity, after the filter has reached constant weight at a temperature of 20°C and relative humidity of 50%, when the filter is placed at the center of the weighing plate. The weight should be corrected by the extrapolation method when the weighing temperature and/or humidity differ greatly from the constant weight conditions. Since membrane filters are sometimes statically charged, weighing the filter on a small weighing plate may produce errors.

ii) Collection of particulate matter

(1) Check the grading unit for dirt.

(2) Check the sampler to confirm that it operates normally.

(3) Fix the weighed filter to the filter holder so that all leaks are prevented. For collecting particulate matter for metal ingredient analysis, separate the collecting filter from the net by inserting a nylon net or fluorine resin filter with a low pressure loss between the filter and metal net.

(4) Switch the power source on and record the starting time of collection.

(5) Adjust the flow meter float to the specified flow rate graduation.

(6) After approximately 5 min of collection, measure the differential pressure with the vacuum meter (or manometer) shown in Fig. 2.34 for correction of the suction flow rate. Adjust the float position to the graduation correctly representing the specified suction flow rate.

(7) Check the suction flow rate at least once a day and measure the differential pressure each time to adjust the float position to the graduation correctly representing the specified suction flow rate.

(8) Record the time of ending collection and calculate the sucked air flow.

iii) Weighing of filter after collection

Weigh the filter after allowing it to stand for at least 24 h under the same conditions as in i).

iv) Data to be recorded

Record the sampling place and date, filter number, time at the start and end of sampling, useful meteorological data (weather condition, temperature, humidity, wind direction, wind velocity, etc.) and collector's name for each collected sample.

d. Calculation of the suspended particulate matter concentration

The suspended particulate matter concentration in $\mu g/m^3$ is calculated from the filter weight and suction flow rate using the following equation.

$$\begin{matrix}\text{Suspended particulate}\\ \text{matter concentration}\end{matrix} = \frac{W_e - W_s}{V} \times 10^3, \qquad (2.6)$$

where W_e is the filter weight after sampling (mg), W_S the filter weight before sampling (mg), and V the suction air flow rate (m^3).

e. Filter division

The filter is divided along the central line if required for measuring or analytical purposes.

f. Flow rate correction

Measurement of the pressure loss arising from the air resistance at the filter or in the sampler system and calibration of the flow meter for the purpose of constantly maintaining the specified suction flow rate, are required when collecting dust with a low volume air sampler. The necessary flow meter calibration for this purpose is described in this section.

Several types of low volume air samplers with different operating mechanisms are available. Since it is impractical to have separate flow rate correcting methods for each type, the correcting method for the simplest collecting system will be described here. The construction of this system is shown above in Fig. 2.34. The filter holder is connected to the rotor meter through a T-shaped tube. A vacuum meter or mercury manometer is connected to the free end of this T-shaped tube. The vacuum meter or manometer is capable of measuring up to 200 mm Hg, but a wider measuring range is sometimes required depending on the kind or diameter of the filter.

i) Principle of pressure correction of the flow meter (rotor meter) indication

Let Q_r denote the rotor meter reading and Q_0 the flow rate (1/min) at 1 atm. Then,

$$Q_0 = C_p \cdot Q_r, \tag{2.7}$$

where C_p is the pressure correction coefficient derived from the following equation.

$$C_p = \sqrt{\frac{P}{P_0}}, \tag{2.8}$$

where P_0 is the design pressure (ordinarily 1 atm) of the rotor meter and P is the pressure within the system under the operating conditions.

Putting ΔP (mm Hg) as the pressure loss within the system measured by the manometer, and $P_0 = 760$ mm Hg, then Eq. (2.8) becomes

$$C_p = \sqrt{\frac{760 - \Delta P}{760}} \tag{2.9}$$

If the low volume air sampler sucks air at a flow rate of $Q_0 = 20$ 1/min, then Q_r becomes

$$Q_r = 20\sqrt{\frac{760}{760 - \Delta P}} \tag{2.10}$$

Thus, the flow meter reading (float position) is set on the basis of the calculated value of Q_r obtained from Eq. (2.10).

Fig. 2.36, the flow rate setting curve, shows the relation between ΔP and Q_r represented in Eq. (2.10), and can be used to correct the flow rate indication.

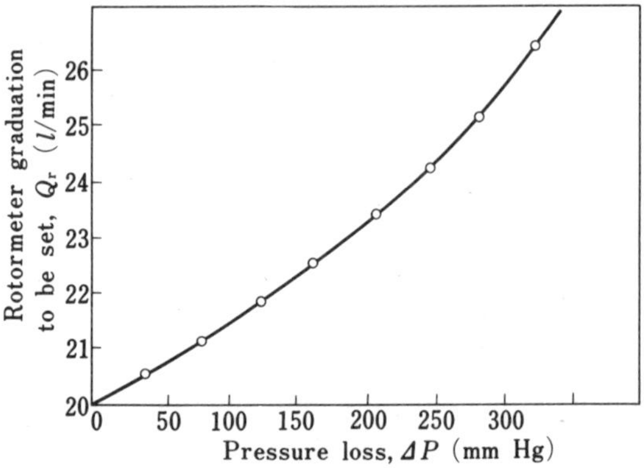

Fig. 2.36. Flow rate setting curve for $Q_0 = 20$ 1/min.

ii) Procedure

(1) Leak test

a) Tightly shut the front opening or all the openings of the filter holder.

b) Close the vacuum meter cock (if not closed, the vacuum meter may be damaged) and check that the rotor meter float indicates zero during pump operation.

(2) Calibrating procedure. Immediately after starting sampling and during sampling, the following procedure should be adopted to maintain the set flow rate.

a) Derive the flow rate to be set (or rotor meter position, Q_r) from the vacuum meter reading (ΔP) using the flow rate setting curve (Fig. 2.36) and adjust the rotor position to the derived flow rate by regulating the pump by-pass valve.

b) Read the gradual pressure drop (ΔP) with change in rotor position and derive the flow rate to be set (Q_r) from Fig. 2.37.

c) Adjust the rotor position to Q_r by regulating the pump by-pass valve. Repeat the above procedure until P reaches a constant value.

g. Notes

(1) The blade of the suction pump should be renewed after about 1 yr (*ca*. 8000 h of service).

(2) Since many flow meters are designed for use at 20°C, temperature correction is generally not a significant factor. (Less than ± 2% error for a ± 10°C temperature difference usually requires no correction.) However, care should be taken since the flow meters may have graduations on a 0°C or 20°C basis dependent on type (maker or standards).

(3) Unit cleaning, etc. (see Table 2.7)

TABLE 2.7. Unit cleaning

Unit	Frequency	Method of cleaning
Packing	Every sampling	Clean with synthetic detergent, etc. However, the use of an ultrasonic cleaner is more efficient and effective.
Net	Every sampling	
Flow meter	Once a year	Clean with synthetic detergent, etc.
Pump silencer	Once a year	Replace felt filter.

(4) Since flow rate change increases with increase in temperature and collected dust, the flow rate should be checked without fail when the humidity increases due to rainfall, etc. Since the filter gradually becomes clogged with dust, the flow rate should be checked more frequently in the latter half of long-term sampling. To this end, the sampler should be installed in a position readily accessible to flow rate measurement or regulation.

(5) Since many rotor meters are designed for use at 20°C, temperature correction is generally not a significant factor. (Less than ± 2% error for ± a 10°C temperature difference usually requires no correction.) However, care should be taken since the flow meters may have graduations on a 0°C or 20°C basis dependent on type (maker or standards).

1. Practice in Japan for measuring the concentration of suspended particulate matter

Since the Japanese environmental criterion for "suspended particulate matter" is based on a size of not more than 10 μm, an apparatus having a maximum permeable particulate size of 10 μm and a permeability of 50% for 8 μm particles at a constant filtering velocity of 20 1/min is used. (Fig. 2.37 shows the overall collecting characteristics.) Environmental Agency, Japan thus specifies the use of a sampler fitted with one or other of the following size distribution units.

(1) Cyclone size distribution unit. The cyclone is designed to remove particles greater than 10 μm in size. An example of the size distribution unit is shown in Fig. 2.38, and an example of a sampler fitted with such a size distribution unit is shown in Fig. 2.39.

(2) Elutriator type size distribution unit. This size distribution unit consists of many thin horizontal plates superposed on one another at narrow intervals. Fine particles in the sampling air pass through the size distribution unit, but those larger than 10 μm settle on the plates (see Fig. 2.40).

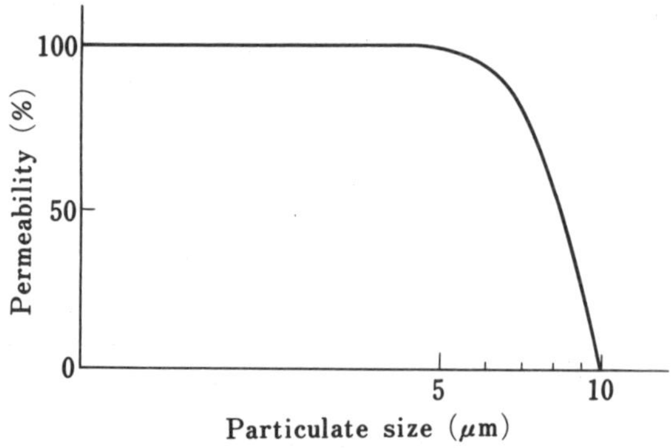

Fig. 2.37. Collecting characteristics curve of apparatus for measuring "suspended particulate matter".

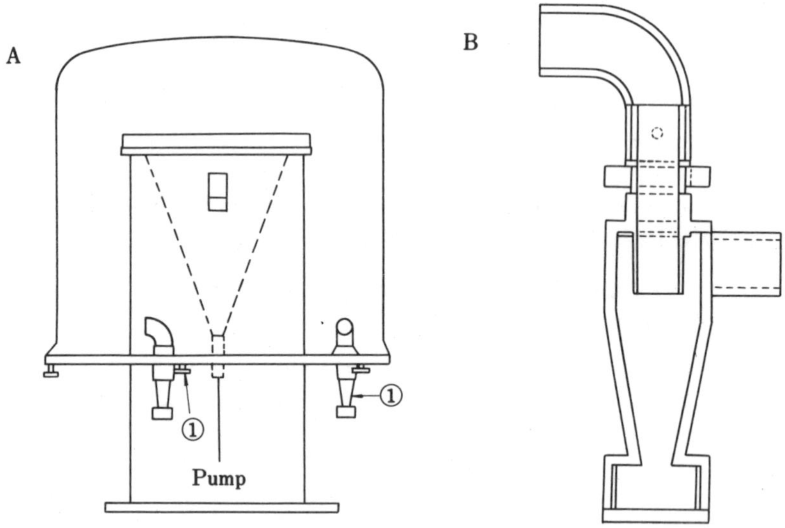

Fig. 2.38. Cyclone size distribution unit. B gives an enlarged cross section of the size distribution cyclone ① in A.

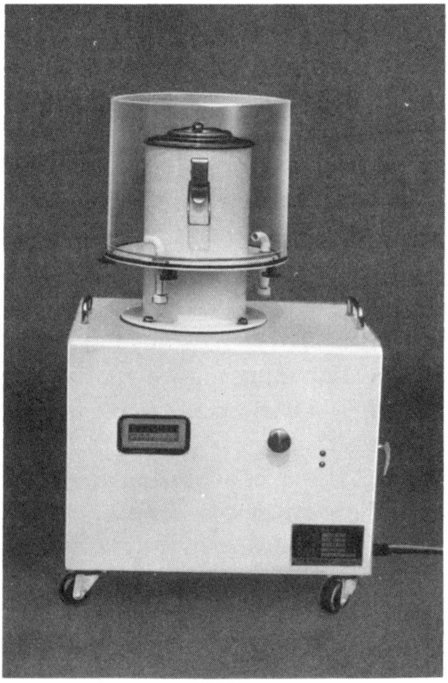

Fig. 2.39. Low volume air sampler cyclone size distribution unit for collecting at 20 1/min (Shintaku Kikai Seisakusho Co.).

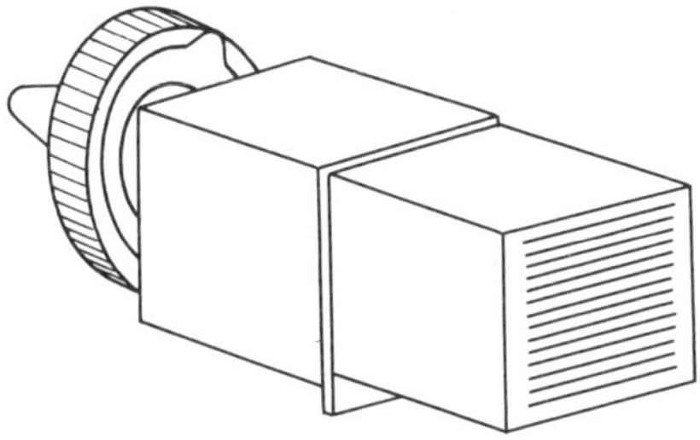

Fig. 2.40 Elutriator type size distribution unit.

C. Particle size distribution analysis

As discussed in Chapter 1, the concentration, chemical composition and size distribution of all suspended particulate matter should be determined to assess its effects on human health. Although particulates larger than 10 μm settle well, remaining in the air for only short periods of time, and enter the respiratory system at a low rate, finer particulates suspended in the air enter deeply into the lung via the trachea and bronchi and exert lasting effects on lung function. Recently, the human effects of lead discharged from automobiles and cadmium from zinc refineries have been discussed.

These metals exist mainly in fine atmospheric particulates not larger than 0.5 μm in size. Growing attention has thus been paid recently to size distribution analysis. Measurements with Andersen samplers or cascade centripetal samplers have rapidly spread and sometimes metal composition analysis is practiced in combination. Analysts should thus be familiar with the principles, characteristics and proper operation of such samplers.

a. Andersen sampler

i) Outline of principle

The Andersen sampler (Fig. 2.41) devised by Dr. A.A. Andersen usually consists of 8 anticorrosive aluminum alloy stages superposed one above the other. Each stage has 200–400 jet openings and a stainless steel or glass collecting disk at the base.

The jet flow velocity of the sampler air during sucking with the constant-feed vacuum pump from the sampler air inlet port operating at a constant flow rate (1 cfm = 28.3 l/min) increases in the lower stages of the sampler. The size distribution of particulates collecting at each stage is shown in Table 2.8.

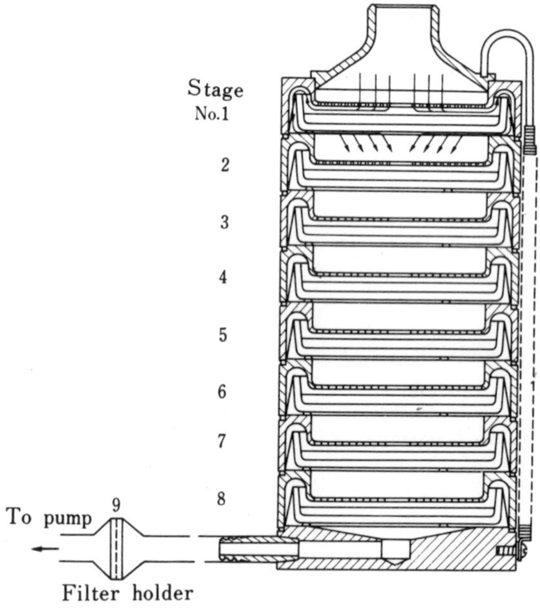

Fig. 2.41. Cross sectional view of an Andersen sampler.

TABLE 2.8. Stage distribution of particulate sizes (Dp 50) with an Andersen sampler

Stage	Particulate size (Dp 50)
No. 1	>11 μm
2	11
3	7.0
4	4.7
5	3.3
6	2.1
7	1.1
8	0.65
Backup filter	<0.43

ii) Theoretical equation

The impacting efficiency (η) of an impactor is defined as the ratio of the cross-sectional area of the jet flow from which particulates of a given size are separated to the total cross-sectional area of the jet flow. According to the particulate jet collision analysis of Ranz and Wong *et al.*, the relation between the efficiency η and the dimensionless inertial collision parameter ψ, i.e. the ratio of stopping distance to jet opening diameter, can be expressed by the following equation.

$$\psi = \frac{C\rho U_0 d_e^2}{18\,\mu\omega}\begin{Bmatrix}\psi_{50}\ (\text{efficiency}(\eta)\ 50\,\%) = 0.14\\ \psi_{95}\ (\text{efficiency}(\eta)\ 95\,\%) = 0.28\end{Bmatrix}, \tag{2.11}$$

where C is the experimental correction coefficient, $1.00 + 0.16 \times 10^{-4}/d_e$ for air at normal temperature and pressure, d_e is the effective diameter of the particulates to be collected (cm), μ the air viscosity (poise), ω the circular jet opening diameter (cm), ρ the particulate density (gm/cm^3), and U_0 the aerosol jet velocity (cm/sec.) Then, rearranging Eq. (2.11), the effective diameter of the particulates to be collected is given by $d_e = \sqrt{18\mu\omega\psi}/\sqrt{C\,\rho U_0}$.

iii) Estimation of the Andersen sampler with standard particles

The author's group has estimated the size distribution characteristics of the Andersen sampler with monodispersed lead fumes and confirmed the characteristics shown in Fig. 2.42. An outline of this study is given next.

EXPERIMENTAL DETAILS. Lead fumes with a prescribed average particle size were continuously fed into a cylindrical aerosol chamber with a capacity of approximately 0.5 m^3. They were then sucked from the chamber into the Andersen sampler at a rate of 28.3 1/min and the collected lead fumes were dissolved in 10% nitric acid and determined by

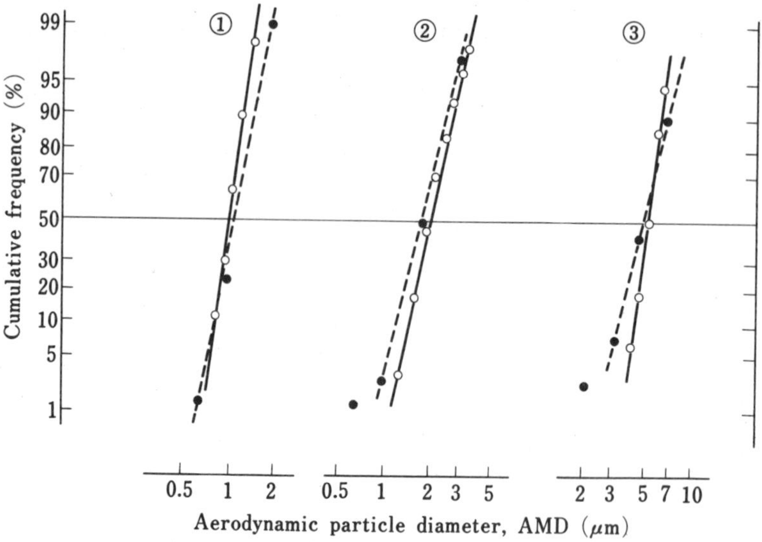

	AMD (μm)
①	
Andersen sampler	1.2
Electron microscope	1.1
②	
Andersen sampler	1.9
Electron microscope	2.1
③	
Andersen sampler	5.0
Electron microscope	5.2

Fig. 2.42. Estimation of the Andersen sampler with standard particles (test aerosol: Pb fumes, $\rho = 11.34\ g/m^3$). ●, Particle size distribution by the Andersen sampler; ○, particle size distribution obtained by the electron microscope method.

atomic absorption spectrometry. The determined weight was plotted against the specified collecting size of each stage for the purpose of preparing a size distribution diagram. However, the reference lead fume particle size distribution was determined on the basis of the aerodynamic particle diameter (AMD) converted from the count median diameter (CMD) obtained from electron microscope photographs of the lead fumes collected by the electron microscope method. The sampler estimation was made by comparing the aerodynamic particulate diameter so obtained with the particulate size distribution on the sampler.

RESULTS AND DISCUSSION. The ADM on the Andersen sampler agreed with that obtained by the electron microscope method. However, the standard deviation fluctuated appreciably. The reason appears to be that some re-dispersion of particles once settled on the collecting disk, nozzle clogging, particle adhesion on the nozzle plate back, and particle settling around the boundary between the collecting disk and succeeding nozzle plate occurred at particular stages.

Special care should be taken to maintain the flow rate truly and constantly at 1 cfm (28.3 l/min) during sampling. Since many marketed samplers are not equipped with a constant-flow mechanism, a differential pressure gauge should be provided for the maintenance of constant flow.

Membrane filters with a 1 μm pore size are recommended for the back-up filter inserted for chemical analysis after the final stage. Cellulose filters are unsuitable for long-term sampling due to their high hygroscopicity, which leads a rapid increase in pressure loss. However, Microsorban filters may be used.

Placing a 0.5 mm thick glass micro-plate on a 1.2 mm thick Tenpax hard glass plate permits relatively precise weight concentration measurements when measuring only the weight concentration by stage. A 0.3 mm thick micro-plate may be used with a 1.5 mm-thick Tenpax glass plate, but special care should be paid taken since the 0.3 mm thick micro-plate is liable to break easily. A five-bit balance (with a reading limit of 0.01 mg) should be used. A high-grade hard glass plate (preferably made of Tenpax glass) should be placed on the sampling disk when sampling for metal ingredient analysis using atomic absorption spectrometer. The sampling period should be 3–5 consecutive days and decided according to the detection sensitivity of the atomic absorption spectrometer. After sampling, the glass plate is placed in a crystallizing disk to which 1:10 nitric acid is added. After ultrasomic cleaning for 1–2 min, a small amount of hydrogen peroxide solution is added for heat extraction.

Although the standard Andersen sampler is designed for use at a sucking rate of 1 cfm, Andersen samplers of high volume type (1000 l/min) are also available.

iv) Example of practical applications

The effect of rainfall on the size distribution of particulate matter containing metal elements has been investigated with an Andersen sampler by Hashimoto *et al.*[15] of the Engineering Department, Keio University. The results are shown in Fig. 2.43. The investigations were conducted from August, 1975 to March, 1976 on the Keio University campus at Hiyoshi, Yokohama City. The results indicated that the amounts of larger particles were considerably reduced by rainfall but those around 0.5 μm in size were not markedly reduced. The iron distributed in larger particles was markedly

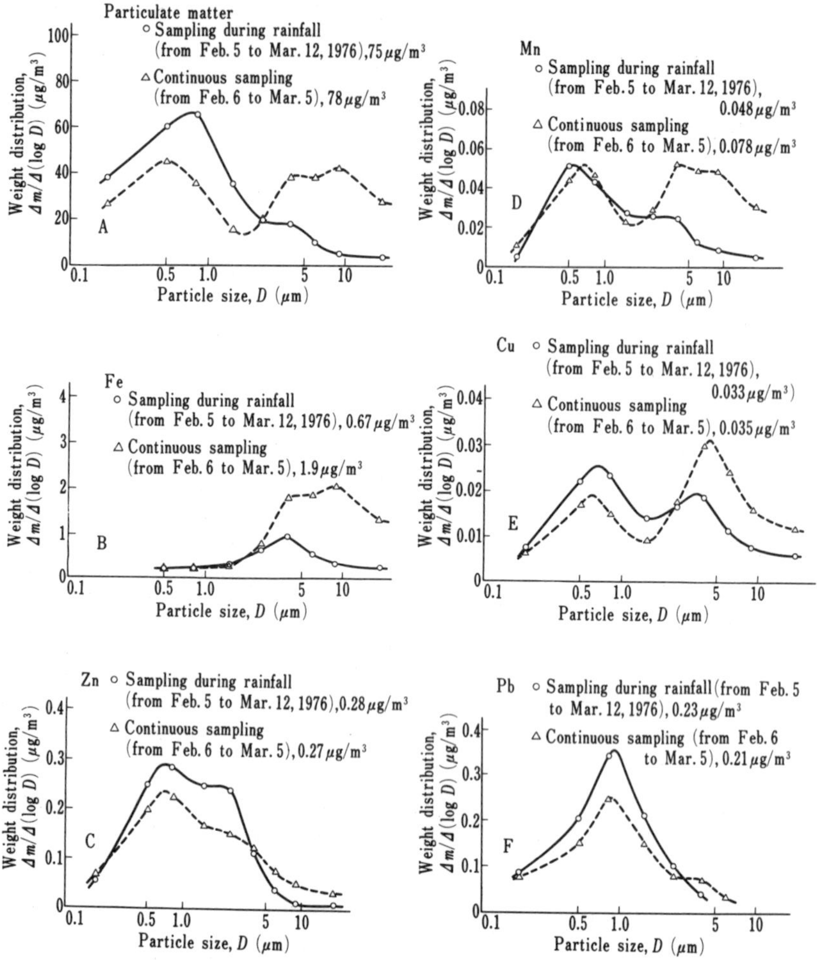

Fig. 2.43. Effect of rainfall on particle size distribution.

decreased by rainfall, but the lead distributed in smaller particles was negligibly decreased.

b. Cascade impactor

An obstacle in the path of an air flow containing particulate matter changes the flow direction of the air. The particulate matter is diverted from the flow direction by inertial forces and collides with the obstacle. Cascade impactors consist of not less than 2 stages of mechanical devices arranged in series for separating and collecting particulate matter by the action of inertial forces (see Fig. 2.44). Since the nozzle size becomes smaller in the lower stages, the velocity of flow increases in the lower stages even at the basic same flow rate. The impactor thus collects larger particles in the upper stages and smaller ones in the lower stages. A mechanical outline of the apparatus is given in Fig. 2.45. It consists of the cascade body with 6 stages of nozzle-collecting plate combinations and a backup filter arranged in series along the air flow, a gauge for measuring the pressure at each nozzle, a vacuum pump for sucking the air, and a flow meter.

2.3. Dust fall

Dust fall is generally defined as those particles among atmospheric particulate pollutants including soot and dust which are not smaller than

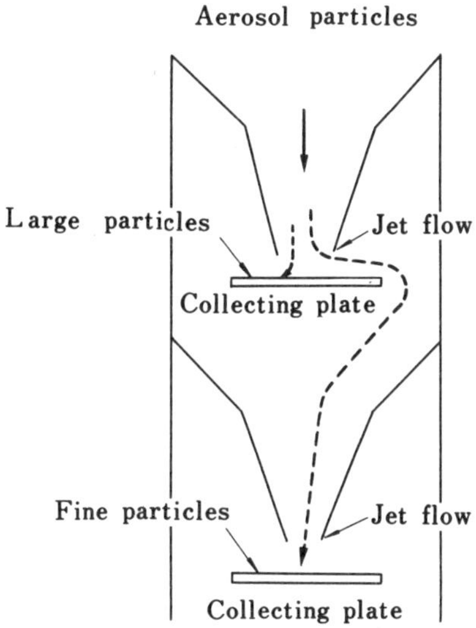

Fig. 2.44. Impactor collection principle.

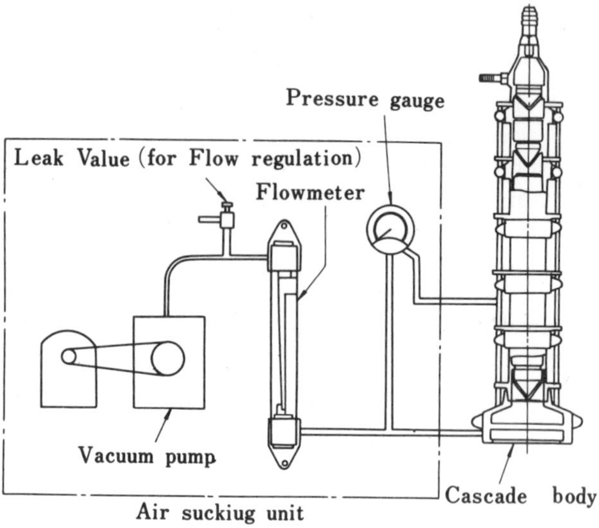

Fig. 2.45. Mechanical outline of the cascade impactor.

20 μm and are large enough to settle by gravity or rainfall. Dust fall is generally measured on a monthly basis with a dust jar or other equipment. The measurements, expressed in $t/km^2/30$ days or mg/cm^2 30 days, indicate the amount of pollutant deposit over a certain area but not the amount of pollutant deposit from any particular discharging source.

A. Dust fall meter

The deposit gauge of the British Standards is the most widely used item of equipment, but the dust jar adopted in the U.S.A. and the simple dust meter adopted in the Hygienic Examination Method of the Japan Pharmaceutical Society are also available.

The deposit gauge of the British Standards (see Fig. 2.46) consists of a collecting bottle (10 to 20 1), hard glass collecting funnel (approx. 30 cm in diameter), bird screening net, inverted funnel, rubber tube and metal stand. A 20 1 collecting bottle is recommended for use in rainy areas, since the collecting surface of the gauge is always dry and collected dust is liable to escape. In addition, the gauge is easily broken and difficult to transport since it is made of glass.

Dust jars with a polyethylene collecting container are widely used in the U.S.A. and have recently been in Japan. According to the American standard method,[16] the dust jar is a cylinder 15–20 cm in diameter and 27

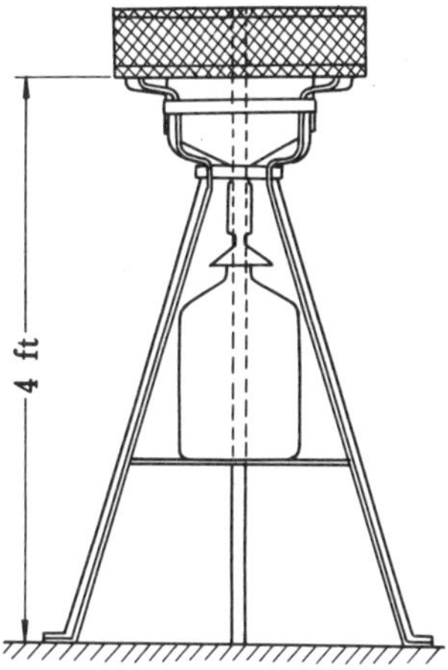

Fig. 2.46. Deposit gauge of the British Standards.

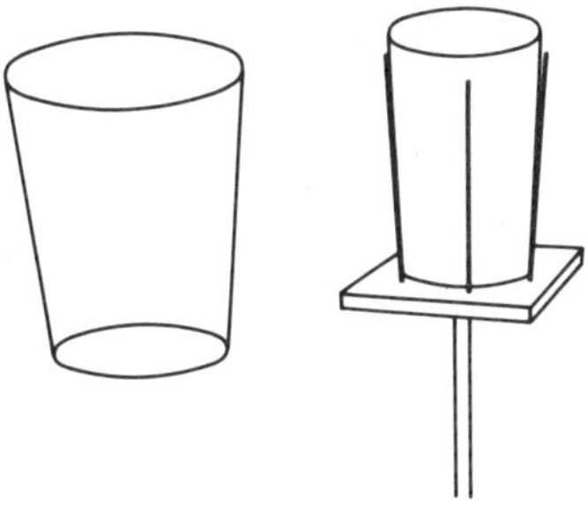

Fig. 2.47. Dust jar.

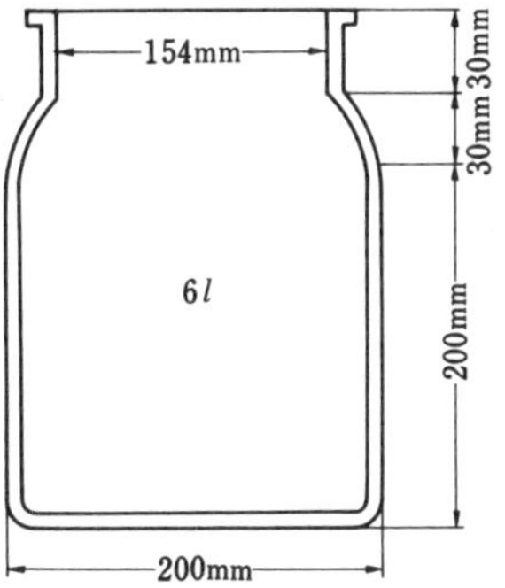

Fig. 2.48. Simple dust meter (wide-mouthed bottle).

cm in length (see Fig. 2.47). The main advantages of this dust jar are (1) ease of handling, (2) ease of transport, (3) robustness, (4) low cost, and (5) no dust escape of the type encountered with the deposit gauge of the British Standards. Based on these advantages, such dust jars are gradually replacing the deposit gauge.

The simple dust meter adopted in the Hygienic Examination Method of the Japan Pharmaceutical Society comprises a wide-mouthed glass bottle (see Fig. 2.48). It is also a kind of dust jar.

a. Comparison of the relative collecting efficiency of the deposit gauge, dust jar and simple dust meter

The collecting efficiency of the deposit gauge, dust jar and simple dust meter was studied on the basis of dust fall measurements by the Japan Health and Welfare Ministry in 1968. The results indicated that the dust jar and simple dust meter have 40–70% higher collecting efficiency than the deposit gauge. The main reason was the difference in degree of escape of collected dust.

The purpose of dust fall measurements is to clarify the general air pollution conditions and to provide data for annual pollution trend analysis and areal comparisons. For these purposes, closer, wider and more frequent measurements are desirable. Sometimes, the scope of measurement covers several areas and large amounts of measuring material are required. The dust jar is superior to the deposit jar due to its ease of transport and analysis, in addition to the higher collecting efficiency.

b. Notes on dust fall meter installation and sampling

Dust fall meters should be installed under the following conditions.

(1) Install the meter away from any local discharging source. If a chimney occurs in the vicinity, install the meter at least 10 chimney heights away from the chimney.

(2) Installation at an elevation from 5–17 m above the ground is preferable in order to minimize the entry of sandy dust from the ground.

Dust fall should be sampled under the following conditions.

(1) Add 10 ml of 0.02 N copper sulfate or 1 g of *p*-chlorophenol to prevent algal growth within the dust fall meter. In the U.S.A., addition of 1–2 ppm of quaternary ammonium salt (the APCA method) is practiced. *P*-Chlorophenol and quaternary ammonium salt are more advantageous than copper sulfate when sampling for analysis of the metal content of dust fall.

(2) Add isopropyl alcohol or ethylene glycol as an antifreezing agent.

(3) Continue the sampling for 30 days and recover the sample monthly.

Carefully wash the inside of the funnel and connecting tube with distilled water or the supernatant in the collecting bottle. Remove large dust particles with a 20-mesh net.

B. Dust fall analysis

Dust fall should be tested for the following items.

(1) pH. This should be measured with a pH meter as soon as possible after sampling.

(2) Rainfall (collected liquid quantity). This should be measured after removing large dust particles with a 20-mesh stainless steel or polyvinylidene chloride sieve.

(3) Dissolved and undissolved matter. The collected liquid may be filtered with a Gooch's crucible or Whatman No. 41 filter. However, fine particles may pass through the filter, and Gooch's crucible should not be used for metal ingredient analysis since the asbestos of the filter material contains high levels of contaminant. Application of a filtering apparatus exclusively using a membrane filter with a low content of metals is thus recommended. The residue on the filter is dried at 105°C for 3 h, allowed to cool in a desiccator to constant weight, and then weighed. This weight represents the total amount of undissolved matter.

A fraction (1–2 l) of the filtrate is transferred to a beaker and concentrated to not more than 100 ml by evaporation. The concentrate is transferred to a platinum or porcelain evaporating dish for further concentration. When the concentrate has almost solidified, the evaporating dish is dried at 105°C for 2 h, allowed to cool in a desiccator, and weighed. The quantity of dissolved matter in the collected liquid is calculated from the weight difference against the evaporating dish.

The total dust fall quantity (w) is the sum of the amounts of dissolved and undissolved matter, and the dust fall quantity is calculated from w using the following equation.

$$\text{Dust fall quantity} = 1.273 \times \frac{w}{D^2} \times \frac{30}{n} \times 10^4 \ (t/\text{km}^2/30 \text{ days}), \quad (2.12)$$

where w is the total dust in the collecting container, D the dust meter funnel diameter (cm), and n the number of days of collection.

(4) Ash content. The dissolved and undissolved matter is ashed at 800°C in an electric muffle furnace. This temperature is proper only for determining the ash content, not for determining the metal content in the ash since some metal compounds are dissipated and insoluble matter forms as a result of fusion of silicate with phospate at a high temperature such as 800°C.

(5) Metal content. There are two pretreatment methods: acid elution after ashing and direct acidolysis or alkali melting after drying. Ashing at 500°C (not 800°C) may be adopted but still gives rise to various problems. Application of a low-temperature ashing apparatus is most recommended. After ashing, the sample is transferred to a 100 ml egg-shaped flask and 20% (or 6 N) hydrochloric acid is added. After heating for 2 h on a water bath with a reflux condenser, the sample is filtered through a Whatman No. 41 or 43 filter. The residue on the filter is then re-eluted with 20% hydrochloric acid. The residue on the filter is transferred to a crucible, ashed after drying in a drier and transferred to a platinum crucible. A small quantity of conc. hydrochloric acid and subsequently the same quantity of hydrofluoric acid are added. The mixture is heated to a solid residue which is dissolved in 20% hydrochloric acid for filtering. The filtrate obtained is combined with the previously prepared filtrate to serve as the test solution.

The hydrochloric acid elution residue of undissolved matter in dust fall may be directly treated with hydrofluoric acid. Iki *et al.* have developed an alkali melting method using the following procedures. The sample is accurately weighed and then transferred to a platinum crucible. A mixture of sodium and potasssium carbonate is added and the sample is carefully melted. Acidification with hydrochloric acid and evaporation to a solid residue are repeated until silicic acid is completely converted to anhydride. The sample is filtered and the filtrate is diluted to a prescribed volume. The diluted sample prepared by the above-mentioned pretreatment method serves for determination of the metal content by atomic absorption spectrometry, emission spectrometry or colorimetry.

References

1. A.L. Benson *et al.*, *J. Air Pollution Contr. Ass.*, **25**, 274 (1975).
2. C. Gelman *et al., Am. Ind. Hyg. Ass. J.*, **36**, 512 (1975).
3. D. A. Lundgren *et al.*, *ibid.*, **36**, 866 (1975).
4. K. Homma, *J. Ind. Hyg. Japan* (Japanese), **13**, 1 (1973).
5. K. Homma, *ibid.*, **16**, 1 (1976).
6. T. Tanaka *et al.*, *Bunseki Kiki* (Japanese), **12**, 347 (1974).
7. R. Dams *et al.*, *Environ. Sci. Technol.*, **6**, 441 (1972).
8. T. Otoshi, *personal communication.*
9. K. Oikawa *et al.*, *unpublished.*
10. Environment Agency, Japan, *Air Pollutant Simple Measuring Technique Research Report* (1973).

11. M. Tani and J. Y. Hwang, *Bunseki Kiki* (Japanese), **9**, 618 (1972).
12. L. C. Lindeken, R. L. Mordin and K. F. Petrock, *Health Phys.*, **9**, 305 (1963).
13. G. A. Jutze and K. B. Foster, *J. Air Pollution Contr. Ass.*, **17** (1), 17 (1967).
14. U. S. EPA Standard Method, *Federal Register*, **36**, no. 84, April 30 (1971).
15. H. Kobayashi and Y. Hashimoto, *J. Japan. Soc. Air Pollution*, **11**, 81 (1976).
16. TR-2 Air Pollution Measurement Committee, Recommended Standard Method for Continuing Dust Fall Survey (APM-1, Revision I), *J. Air Pollution Contr. Ass.*, **16** (7), 372 (1966).

CHAPTER 3

SAMPLE PRETREATMENT AND PREPARATION FOR ANALYSIS

3.1. Preparation of the sample solution

A. Sample division (cutting)

After weighing sample material acquired from a high volume sampler, division is conducted in the manner shown in Fig. 3.1. However, it is not necessary to follow such a division formally, and the cut proportions may be increased or reduced according to actual circumstances or in compliance with the objective components for analysis.

In the analysis of metallic components, an ordinary 2 × 7 in sector (about 22 % of the sample) is used, and after ashing, acid extraction is carried out. The concentration of metallic components is measured by atomic absorption spectrometry or emission spectrographic analysis. Since chrome is essentially undissolved by extraction with hydrochloric or nitric acid, another 2 × 7 or 1 × 7 in sector must be taken, and treated by alkali fusion or some other similar operation.

In the determination of the sulfate and nitrate ion fractions, a 7 × 3/4 in (8 %) sector is used. For determining organic substances, a 2 × 7 in (22 %) sector is taken and extracted with benzene using a soxlet extraction apparatus and the results are given as the total amount of benzene soluble organic substances. In some cases after concentration of the extracted solution, extractive separation is conducted using cyclohexane or nitromethane, and identification and quantitation of various organic components are conducted by gas chromatography or infrared spectroscopy.

B. Pretreatment of the sample

Among the important factors influencing the accuracy of chemical analysis are the sampling technique and pretreatment. Regardless of a high degree of analytical reliability and the employment of analytical equipment with high sensitivity, the value of such technical and instrumental quality will be lost if insufficient consideration is given to the sampling methods and pretreatment. Many reports on the analysis of atmospheric suspended particulate matter exist; however, uniform and standardized methods are

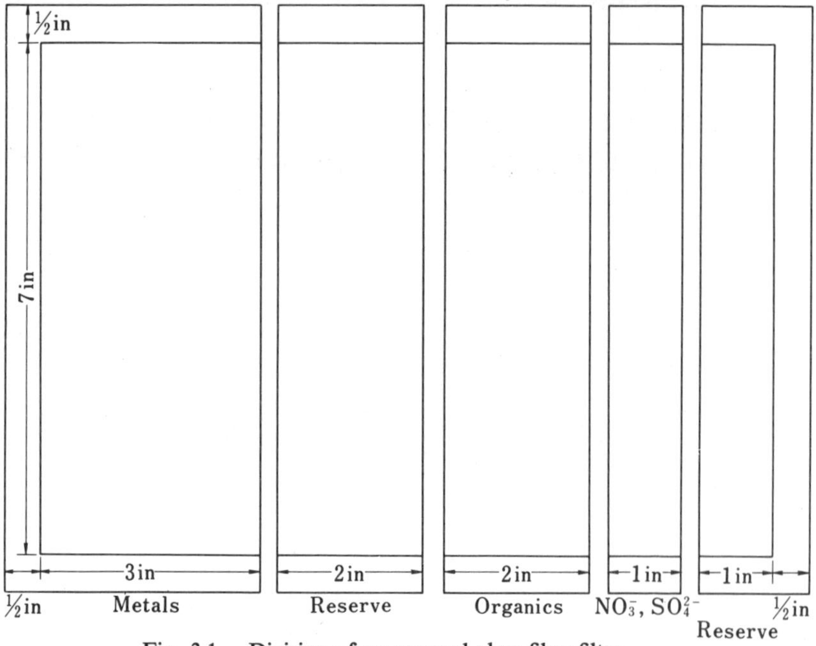

Fig. 3.1. Division of an exposed glass fiber filter.

relatively few, and individual analysts currently employ a variety of techniques. However, in Japan, legislation for an official analytical method coded JIS K 0097–1971, "Determination of cadmium and lead in stack gas", and JIS K 0096–1974, "Determination of chromium and manganese in stack gas", has been enacted, and from now, this will probably be used as a standard method in dust analysis.

a. High temperature ashing (electric muffler ashing)

In high temperature ashing, the dust sample is placed in a crucible and organic substances are destroyed by fire using an electric muffler or burner. The ashing temperature is usually 500–700°C. In comparison with wet oxidative decomposition, the procedure has several advantages; it is simple, the risk of sample contamination by oxidative agents (decomposition agents) is small, and a large number of samples can be efficiently and rapidly treated.

Since it is difficult to regulate and maintain an appropriate ashing temperature with a gas burner, such burners are unsuitable for determining the ash constituents, etc. However, they can be used for evaluating the total ash content.

Many electric muffler ovens have automatic temperature regulators which generally operate at about 500°C. At this temperature, the error is 1% of the gauge reading, and 0.3–0.5% of the apparatus temperature, so that an error of 7–8% must be allowed for in its operation. Also, there is a considerable lack of uniformity in muffler temperature within the muffler itself, of the order of 10% between the front and rear areas. This must be considered when selecting apparatus for high temperature ashing.

The results of an investigation by Hamilton[1] are shown in Fig. 3.2. It can be seen that with the ashing temperature set at about 500°C, a maximum difference of 60°C (12%) was found, dependent on the location within the muffler. Of course, this difference (range) of temperature itself varies according to the make or type of muffler and the size (volume) of the interior. Thus, the temperature indicated on the gauge of automatically regulated ashing ovens does not necessarily show even the average temperature of the interior of the muffler, a point which requires caution.

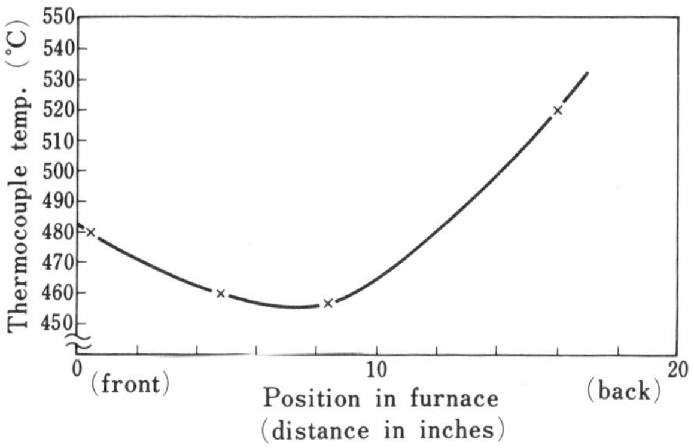

Fig. 3.2. Relationship between temperature and position in a furnace, as determined with standard sentinels.
(Source: ref. 1. Reproduced by kind permission of W. Heffer and Sons, Ltd., England.)

i) Ashing vessel (ashing boat)

For the quantitation of trace metals, quartz boats are considered very practical. Generally, porcelain boats are also extremely useful. Vessels to which oxygen can be adequately supplied such as the shallow bottom type (boat type) are suitable, while crucibles with a deep bottom are not preferred since long ashing times are required.

If porcelain dishes are employed over long periods of time, the glaze is damaged by alkaline ash, and a portion of the ash components becomes fixed in such regions so that it cannot be removed even by hydrochloric or nitric acid treatment. Glazes found to have imperfections should therefore be avoided whenever possible. Melting or dissolution of the glaze, however, very rarely occurs in ashing operations conducted at about 500°C.

ii) Ashing temperature

Samples trapped on glass fiber filters by means of high volume samplers are generally ashed at 500–550°C. However, within this temperature range there are certain substances for which a risk of volatile loss exists during the ashing process. These include Pb, Cd, Sn, As, Ca, etc., which are volatile elements having low boiling points. In addition, there are many compounds which have boiling points below 500°C, such as antimony trichloride, $SbCl_3$ (219°C), molybdenum pentachloride, $MoCl_5$ (268°C), and ferric chloride, $FeCl_3$ (307°C). In circumstances where alkaline phosphates which have low melting points constitute a primary component of the ash, their preferential melting prevents complete melting of the sample, and incompletely combusted carbon which obstructs oxidation envelops it. A somewhat similar problem exists with sodium chloride and potassium chloride which, during the ashing process, form bundles around Cd, Cu, Zn, Fe and many other elements, preventing them from entering solution during subsequent acid treatment. According to Kuzuhara, only 50–60% of Cd, Zn, Cu, etc. were recovered at each concentration tested for both NaCl and KCl, when an electric muffler was employed at a temperature of 500–550°C. Similar results have been found with other metallic elements. On the other hand, in experiments similar to that described above where a low temperature ashing apparatus was employed, recovery rates of nearly 100% have reportedly been obtained regardless of the quantity of sodium or potassium present in the ashing mixture. Food samples and dust collected from coastal regions, where the content of sodium or potassium is relatively high, can thus conceivably display sizable analytical errors resulting from the particular ashing method employed. Also, the possibility of volatile loss must be considered in the case of chlorides since they often have relatively low boiling points.

iii) Adsorption of metals by insoluble substances

When large quantities of silicates are present in ash, these become conjugated with Al, Fe, Mn, etc., and the resulting compounds tend to be insoluble. This situation becomes more conspicuous at higher temperatures, but at ashing temperatures below 500°C, the problem does not arise with Fe, Cu, or Al. Not only are silicate compounds found in glass fiber filters, but they have also been detected in large quantities in atmospheric samples in the form of α-SiO_2 (α-quartz), $Al_2O_3 \cdot 2SiO_2 \cdot 4H_2O$ (endellite), etc. Careful consideration must therefore be given to the possibility of silicate conjugation of metals in analytical work.

Carbon also frequently remains at ashing temperatures below 500°C due to insufficient ashing. When this occurs in samples containing large amounts of phosphates, the carbon exhibits the unique property of selectively adsorbing Cu^{2+}, Pb^{2+}, Cd^{2+}, Fe^{3+}, Co^{2+}, Ni^{2+} and Zn^{2+} ions present in acidic solutions. In particular, carbon shows a strong selectivity for Cu^{2+}, Fe^{3+} and Co^{2+} in solutions acidified with hydrochloric acid, while adsorption does not occur with Mn^{2+}, Al^{3+} or light metal elements. This tendency is also observed in solutions acidified with perchloric acid (Table 3.1).

TABLE 3.1. Effect of carbon on metal recovery (after Kuzuhara)

	Metals	Carbon added (%) per 1 g of ash: 20	10	2	1	0.2	0
		Recovery (%)					
Room temp.	Cu	66.7	75.6	97.8	98.7	100.0	100.0
	Pb	78.8	89.0	93.2	93.2	94.9	95.7
	Cd	94.5	95.0	98.4	98.9	99.5	98.9
500°C (5 min)	Cu	7.7					77.2
	Pb	29.3					82.8
	Cd	61.5					99.0

A compromise exists between the dry and wet ashing methods, in which a small quantity of concentrated nitric acid or concentrated sulfuric acid is added to a porcelain vessel (boat) containing the sample before heating. The ashing time is effectively shortened, and some suppression of carbon adsorption results.

Kometani *et al.*[2)] have conducted experiments for suppressing or preventing losses due to volatilization during ashing and reported good results after sulfuric acid addition. Table 3.2 shows results for recovery rates in the case of high temperature ashing with hydrochloride, nitrate and

TABLE 3.2. Recovery of metal salts (%) from Pt crucibles after heating at various temperatures for 1 h

Metals	μg \ °C	Chloride 400	Chloride 500	Chloride 600	Nitrate 400	Nitrate 500	Nitrate 600	Sulfate 400	Sulfate 500	Sulfate 600
Pb	50	78	54	21	100	102	96	102	102	96
Zn	40	87	87	62	99	97	88	96	97	97
Cd	10	77	56	17	102	100	83	100	100	96
Cu	20	96	96	71	104	98	73	102	100	56

(Source: ref. 2. Reproduced by kind permission of the American Chemical Society, U.S.A.)

sulfate salt of Pb, Zn, Cd and Cu. The sulfate salts showed the best recovery rate (close to 100%) at an ashing temperature of 500°C, followed by the nitrate salts. Poor recovery was seen with the hydrochloride salts, especially those of Pb and Cd. In subsequent experiments, only hydrochloride salt of the four metals or hydrochloride plus added sulfuric acid was evaluated, and good recovery was observed only in the latter case (Table (3.3).

TABLE 3.3. Effect of H_2SO_4 on metal chloride recovery (%) from Pt crucibles after dry ashing, starting at 300°C for 30 min and then at 500°C for 1 h

Metal	μg	In Pt crucible		On paper filter in Pt crucible	
		Chloride	H_2SO_4 added	Chloride	H_2SO_4 added
Pb	50	74	96	85	98
Zn	40	85	100	95	103
Cd	10	76	96	90	96
Cu	20	91	94	87	102

(Source: ref. 2. Reproduced by kind permission of the American Chemical Society, U.S.A.)

The above discussion has covered various problems arising during high temperature ashing. Apart from the immediate purposes of estimating the total ash content, ignition loss, etc., the kinds of metals occurring as atmospheric pollutants are ultimately to be determined. The simple use of a muffler alone in the ashing process is to be discouraged in view of the demerits described, but when it is used in conjunction with the sulfuric acid treatment of Kometani *et al.*,[2)] it becomes a very important piece of equipment in the preparation of sample solutions.

b. Low temperature ashing

As mentioned, the process of high temperature ashing is simple and many samples can be treated rapidly and efficiently, although losses through volatilization, metal adsorption by silicates and carbon, etc. represent problems. Various drawbacks also exist with the wet decomposition method to be discussed below; these include extensive contamination by the oxidants employed, and limitations to the sample quantity which can be handled. Recently, various attempts have been made to overcome these problems. They include the flask combustion method, and the oxygen bomb method, where ashing is performed in a glass chamber wrapped in a high frequency coil into which ozone is introduced.

Around 1960, an elementary low temperature ashing apparatus (Fig. 3.3) was developed for the purpose of removing organic substances by

Fig. 3.3. Low temperature ashing apparatus (IPC 1101B–648AN).

using oxygen and high frequency discharge. Its application was not only for ashing in chemical analysis, but also for eliminating photoresistant substances in silicon wafer semi-conductors, in chemical synthesis, for improving adhesiveness in plastics, etc., and it has since enjoyed extensive practical application. In the field of analytical chemistry, with the appearance of highly efficient multichambers, its utilization has become widely popular.

i) Principle

When a high frequency discharge is conducted in an oxygen atmosphere, oxygen gas becomes excited and forms atomic oxygen. The latter can oxidize the sample, accelerate decomposition, and ashing can be accomplished at a low temperature (room temperature to 150°C). The principle of the apparatus is shown in Fig. 3.4. The sample is first placed within thc reaction chamber (a quartz or Pyrex glass chamber), the chamber exhausted, and oxygen then introduced at low pressure (approx. 1 mm Hg). A high frequency of 13.56 MHz is supplied from an electronic panel attached outside the chamber. The oxygen becomes excited and when a plasma condition has been created, ashing occurs. When only the high frequency and oxygen (containing impurities, nitrogen gas, etc.) are in a plasma condition in the absence of sample, the interior of the chamber exhibits a pink color due to the dark red color of oxygen gas and orange color of nitrogen gas when mixed. When sample is added, there is a color change to faint blue or milky white. This is due to the reaction, $C + O_2^* \rightarrow CO_2$; and CO_2, carbon monoxide, water and free hydroxide radicals are expelled. As ashing proceeds, the chamber becomes purple in color. A pink color then signifies the end point, when ashing in complete Beyond this point, further insertion or addition of sample to the chamber is of no avail

The nature of the reaction mechanism occurring within the reaction chamber can be described as follows. As a result of the high frequency discharge, atomic oxygen, molecular oxygen ion and free electrons in the so-called plasma form at the transmission plate by the following reactions:

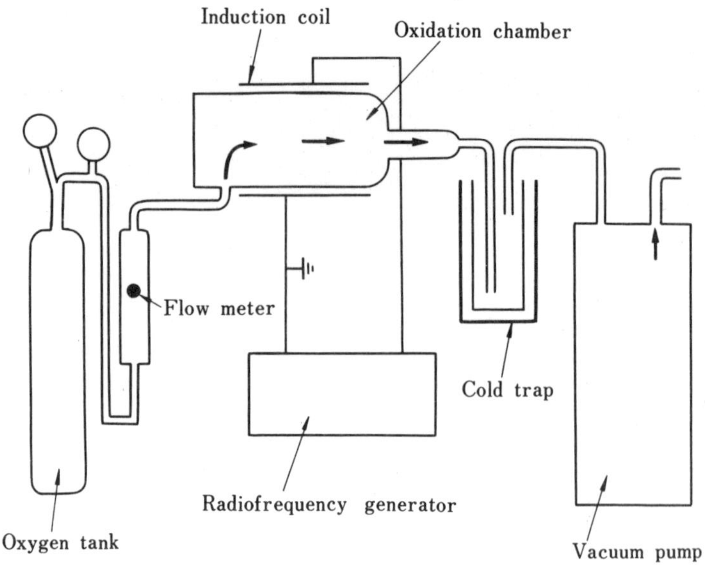

Fig. 3.4. Schematic diagram of a low temperature ashing apparatus.

$$O_2 \rightarrow O\cdot + O\cdot$$
$$O\cdot + O\cdot \rightarrow O_2^*$$
$$O_2^* \rightarrow O_2 + h\nu$$

The most effective conditions are those where the above components occur in the following proportions:

$$O\cdot\ 20\%,\quad O_2^*\ 20\%,\quad O_2\ 59\%,\quad O_1^+ + O_2^+\ 1\%$$

However, the formation of atomic oxygen by direct liberation from molecular oxygen is scanty. Nitrogenous oxidants and atomic nitrogen arise from water and nitrogen gas which exist as impurities in the oxygen used in the ashing process, and it is believed that atomic oxygen arises from a series of chain reactions. That is to say, water and nitrogen play the role of the "spark" when the gas is ignited. It is therefore believed that very little atomic oxygen formation can occur when pure nitrogen is used.

$$NO^+ + e \rightarrow N + O$$
$$N + O_2^+ \rightarrow NO^+ + O$$

ii) Experiments on recovery rate

Table 3.4[3] shows the results of an experiment in which low temperature ashing and that employing a muffler were compared. Dried blood was used as the organic substance for the sample and the labelled metallic inorganic compounds shown in the table were added. It can be seen from the results that even volatile inorganic compounds have recovery rates very near 100% when ashing is done by the low temperature method.

TABLE 3.4. Recovery of radioactive tracers in dry ashing

Nuclide	Sample	Recovery (%)			
		Low temperature ashing		Muffler furnace	
		boat	trap and chambers	400°C, 24 h	900°C, 3 h
^{124}Sb	$SbCl_3$ + blood	99	0	67	9
^{70}As	$HAsO_2$ + blood	100	0	23	0
^{137}Cs	CsCl + blood	100		—	—
^{60}Co	$CoCl_2$ + blood	102	0	98	30
^{67}Cu	$CuCl_2$ + blood	101	0	100	58
^{51}Cr	$CrCl_3$ + blood	100	0	99	56
^{198}Au	$AuCl_3$ + blood	70	30	19	0
^{50}Fe	$FeCl_3$ + blood	101	0	86	27
^{210}Pb	$Pb(NO_3)_2$ + blood	100	0	103	13
^{54}Mn	$MnCl_2$ + blood	99	0	99	79
^{202}Hg	$Hg(NO_3)_2$ + blood	92	8	<1	0
^{110}Ag	AgCl	72	28	65	21

(Source: ref. 3. Reproduced by kind permission of the American Chemical Society, U.S.A.)

The results of an experiment conducted by the Cincinnati Research Laboratory of NASN (U.S.A.) are shown in Table 3.5. Recovery rates for a number of metals were compared after treatment by muffler furnace ashing or low temperature ashing, using samples of inorganic nitrate compounds applied to fiber glass filters. The results indicated that at a high ashing temperature of about 550°C, losses of Pb, Cd, Zn and Sb exceeded 50%, while the recovery rates were close to 100% at low ashing temperatures. However, standard inorganic compounds were tested in the experiment, and it was assumed that the data closely reflected recovery rates which could be expected for samples of inorganic compounds collected with a high volume sampler. A follow-up experiment by the author which compared recovery rates for high and low temperature ashing, revealed the following (Fig. 3.5): recovery rates were lower for high temperature ashing than low temperature treatment by as much as 45% for Cd, and 33% for Pb and Mg; conspicuous differences showing a similar trend were found in Ni and K.

TABLE 3.5. Effect of ignition method on metal recovery (after G. B. Morg *et al.*)

Metal	Recovery (%)	
	low temperature ashing	muffler furnace (550°C)
Ti	95	92
Cr	112	100
Mn	99	107
Co	96	97
Ni	97	99
Cu	98	92
Zn	96	39
Mo	98	116
Cd	92	53
Sn	95	87
Sb	99	46
Ba	97	99
Pb	101	46

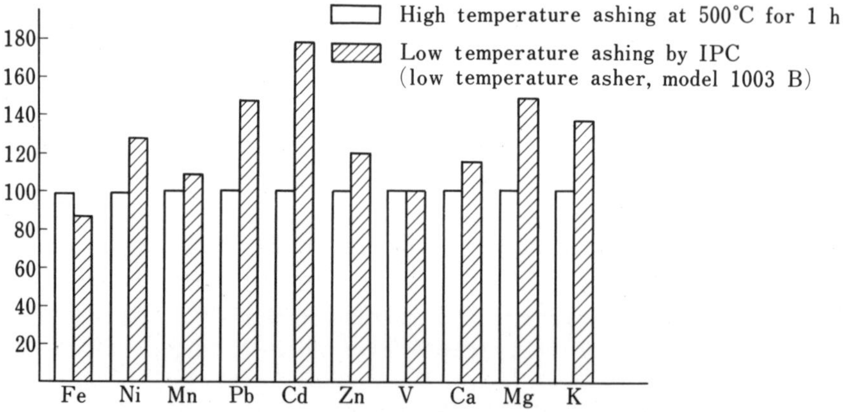

Fig. 3.5. Comparison of AAS results for an urban particulate sample on a glass fiber filter by high temperature ashing and low temperature ashing.

As a result, it was considered that appreciable losses had occurred due to volatilization, fixation or adsorption.

In cases where low temperature ashing is employed, care must be taken over the following points:

(1) The ashing process must be conducted very gently at low or normal pressure in the ashing chamber. Sudden or rapid treatment often causes "bumping" of the sample.

(2) The wattage supply must be reduced when the amount of carbon in the sample is relatively large. If the wattage input remains high in such cases, the chamber temperature becomes very high as a result of heat liberation from the carbon.

(3) Since the interior pressure of the chamber becomes low (0.3–1 torr), the boiling points of metallic compounds within it become much lower than those at atmospheric pressure.

(4) It is believed that the proper degree or extent of oxidation of metallic constituents also represents a problem. The possibility exists that under certain ashing conditions, oxidation progresses too far and the metallic constituents become insoluble under acidic conditions. Furthermore, the author[4] has quantitatively confirmed that recovery rates become poor, as do the ashing time and progress of ashing in instances where a potassium dichromate solution is added to the ashing sample on a cellulose filter. The recovery rate showed a value of about 50% in 2–3 h (Fig. 3.6). It is believed that this can be attributed to the formation of chromium trioxide (Fig. 3.7) and chromium carbonyl.

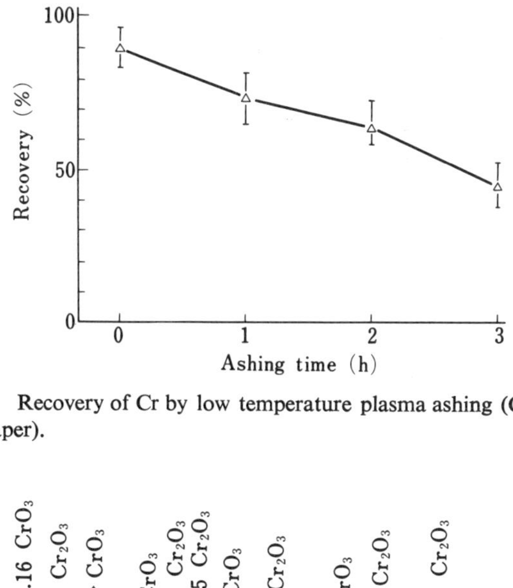

Fig. 3.6. Recovery of Cr by low temperature plasma ashing (Cr added on a filter paper).

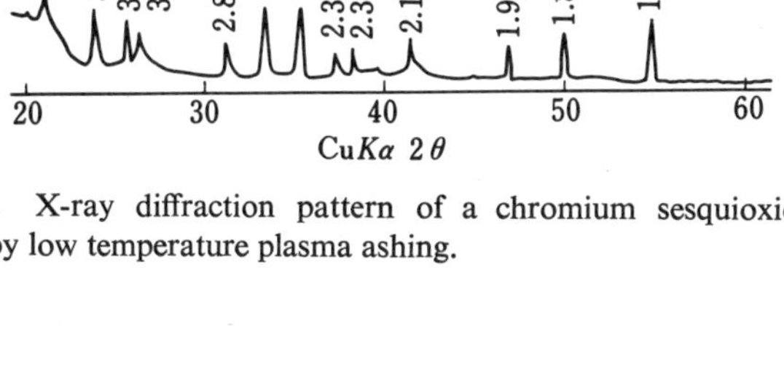

Fig. 3.7. X-ray diffraction pattern of a chromium sesquioxide sample treated by low temperature plasma ashing.

C. Wet oxidative decomposition

Wet oxidative decomposition is a method by which samples are thermally decomposed in some strong oxidizing agent such as sulfuric acid, hydrogen peroxide, perchloric acid, etc. Since decomposition is conducted at low temperatures, evaporative loss of inorganic compounds is lower than in high temperature decomposition. Many combinations of oxidizing agents are commonly used, including sulfuric acid + nitric acid, nitric acid + hydrogen peroxide, nitric acid + perchloric acid, sulfuric acid + nitirc acid + hydrogen peroxide, etc.

Choice of an appropriate oxidizing agent should be made after careful consideration of the sample characteristics, element under study, and characteristics of the oxidizing agent and each acid (Table 3.6). It is recommended that long neck Kjeldahl flasks be used for this operation, installed with reflux condensers in a water bath with heating. Decomposition is completed in 3–4 h if the sample is collected on glass fiber filters. However, attention should be given to the condition of the solution in the flask, for when the color becomes transparent, faintly colored or colorless, the operation is ended; if the sample contains abundant Fe, then the color of the solution becomes light yellow.

As described in section 3.2.A, purification of sulfuric acid is invariably difficult, and appreciable quantities of impurities may be introduced during

TABLE 3.6. Characteristics of decomposition reagents

Reagent	Characteristics
H_2SO_4	Primarily characterized by its dehydration action. In cases where alkaline earth metals are present in large quantities, there is deposition of sulfates. Adsorption of trace metals, particularly lead, occurs. This acid often contains Cd, Ni, Zn, Cu and Pb as impurities. Sulfate ions remain.
Fuming H_2SO_4 Fuming HNO_3	Unsuitable for quantitation in trace analysis since the purity is not good.
HNO_3	Oxidative action. Strong in decomposition of proteins but slightly weak in decomposition of carbohydrates. Decomposition ability of fats is inferior. Not used alone but together with H_2O_2 or $HClO_4$.
$HClO_4$	Powerful oxidative action when heated. Alone, there is a danger of explosion, and it must always be used together with HNO_3. Purification is difficult and it often contains impurities.
H_2O_2	Used as an oxidant in place of HNO_3 or $HClO_4$. Caution should be given to its content of metal impurities.

the oxidation procedure. These include various heavy metals such as Cd, Pb, Zn, Ni, etc., which may be introduced into the sample solution or cause high values in blanks. Also, the nonvolatile characteristic of sulfuric acid gives rise to traces of the acid in samples. Its use for adjusting the pH of the sample solution should be limited or avoided.

In the author's experience, H_2SO_4 and HNO_3, which are used to decompose samples for analysis of atmospheric Cd, give rise to high Cd values in blanks and render analysis impossible. Moreover, when large quantities of aluminum salts are present in the sample, H_2SO_4 reacts with them to form highly insoluble aluminum oxides. On the other hand, any excess calcium in the samples may react with H_2SO_4 to form the very insoluble $CaHSO_4$.

Proper attention must be given to such contamination arising from oxidizing agents, and the use of H_2SO_4 should be avoided entirely if possible. Under circumstances where the sample is collected on glass fiber filters and oxidation by H_2SO_4 is not imperative, HNO_3 and H_2O_2 can adequately suffice as alternatives for this purpose.

D. Fusion decomposition

Cr, Mo, Si, Al and Ti are insoluble in acids and must be traeted by a fusion process for their dissolution. In this method, samples or compounds which are essentially insoluble in acids are placed in a platinum crucible and subjected to a dissolution procedure using a fusion reagent. This converts them to water-or acid-soluble compounds. The fusion reagents include compounds which are alkaline (Na_2CO_3 or K_2CO_3, $CaCO_3$ + NH_4Cl, etc.), acidic ($K_2S_2O_7$), oxidative (Na_2CO_3 + Na_2O_2, Na_2CO_3 + KNO_3, etc.), and reducing (NaOH + KCN). The most suitable fusion reagent must be selected after careful consideration of the characteristics of the elemental compound under investigation.

There have so far been many reports of analytical results for Cr, Mo, etc. in atmospheric particulate material, and all have been conducted with acid extraction as a pretreatment procedure. However, apart from results where the acid extract amount (acid dissolution fraction) of a metal was required, it is believed that in such acid extraction methods, a moderate quantity of similar metals escapes dissolution and extraction, since both the elemental and oxide forms of metals such as chromium and molybdenum are insoluble in oxidative acids. (Results by the author have shown that elution of these metals occurs at rates of only about 20–30%.) The total amount of these elements cannot therefore be given accurately by methods which use acid extraction, and can be determined reliably only by the fusion method. As mentioned above, there are many fusion decomposition methods, various combinations in which fusion reagents can be used, and many possibilities with regard to the composition of each reagent to be used in the fusion mixture. Hence, the selection and final choice of the most suitable fusion system depends upon a critical consideration of the nature of the compound which contains the objective element under study. However, a detailed description of fusion reagent systems considered adequate for each respective element and/or compound would lead to unnecessary confusion, and so only the most frequently employed method is described here. This is the alkaline fusion-sodium carbonate method. The reader is urged to consult appropriate references for details of other methods.

For the purpose of conducting fusion decomposition, it is considered best to collect samples on polystyrene or membrane filters. In particular, major components of atmospheric dust such as Si, Al, Ca, and Mg are also primary conponents of glass fiber filters and are therefore unsuitable. Moreover, since Fe, Cr and Mn are present in relatively large amounts in filters, care must be taken over blank values and errors in values atributable to trace metal contamination, in order avoid serious errors. With filters of the polystyrene or membrane filter type, it is possible to conduct sufficient ashing in low temperature ashing equipment, and the result is a smoothly run analytical preparation, while the level of precision of the analytical results increases due to low levels of contaminants in the blanks.

(1) Case of employing polystyrene or membrane filters in sodium carbonate fusion. A portion of the sample collected on a polystyrene filter (a 2 × 7 inch (22 %) sector is taken in the case of sampling with a high volume sampler) is placed in a borosilicate glass boat (or platinum boat) and ashed in a low temperature ashing apparatus or maffler furnace. Residual ashed substances are transferred to a platinum crucible (20 ml) and after addition of 1–2 g Na_2CO_3, this is sealed. The crucible is placed directly into the fire and starting from a low value, the temperature is slowly raised to about 900°C with occasional shaking. The contents are mixed well for about 20 min under these conditions (heated fusion). On conclusion of the fusion procedure, the crucible while still hot is placed on an iron or polished rock plate and rapidly cooled to loosen the "melt" from the crucible. Upon cooling, a small amount of water is added and when this has been absorbed between the melt and crucible, the crucible is reheated using a small flame from a burner. The melt then separates from it, and is treated by an appropriate separation method (see Fig. 3.8) according to the element analyzed. When Si or Al is the object of quantitation, dissolution is conducted in a water bath. Moreover, when heavy metals are being investigated, dissolution is accomplished in a solution of HCl, and Si and Ti are removed by a separation procedure.

When Si and Al are the objects of determination, the melt is treated by the method shown in Fig. 3.8. The crucible and cover are transferred to a

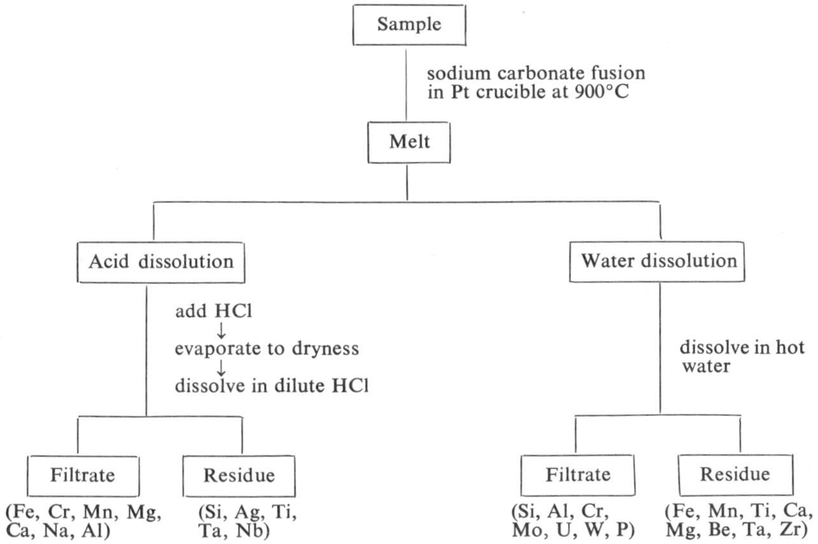

Fig. 3.8. Flow chart for sodium carbonate fusion.

200 ml beaker and the melt dissolved in a small quantity of warm water, Both the crucible and its cover are removed and rinsed with a small quantity of warm water and the rinse is transferred to the beaker. Later, the solution and residues are separated by a filtering operation or by centrifugation. The washing/rinsing operation must be repeated several times to ensure high recovery of the element to be analyzed. The solution contains aqueous components such as Si and Al, while Fe, Mn and Be are present as residual components. Moreover, when Mn and Cr are the heavy metals under consideration, the crucible and filter are transferred to a 200 ml beaker after cooling, and the melt is dissolved in a small quantity of warm water. The crucible and cover are removed and washed thoroughly in a small portion of warm water; the washings are transferred to the former beaker. Next, the beaker is placed on a watch glass and 8 ml of HCl (1 + 1) and 1 drop of H_2O_2 (30%) are added on it. The contents are then boiled, the CO_2 gas expelled, and the melt simultaneously dissolved. The residues are Si, Al and Ti, which are separated by centrifugation at 3000–4000 rpm and washed thoroughly.

(2) Case of employing glass fiber filters. Prior to the alkaline fusion procedure with Na_2CO_3, it is necessary to carry out the following operation. Following ashing of the sample (a 2 × 7 inch (22%) sector) with a low temperature ashing apparatus or maffler furnace, the ash is placed in a platinum crucible. After addition of a few drops of H_2SO_4 (1 + 1) and 5–10 ml

hydrofluoric acid, the mixture is heated on a hot plate until generation of most of the white smoke liberated from H_2SO_4 has stopped. Care must be taken to ensure that the platinum crucible is heated quietly and that the temperature is increased very slowly. After cooling, the Na_2CO_3 fusion is conducted according to method (1) above.

E. Examples of sample solution preparation

A few examples of sample solution preparation focused on atomic absorption spectroscopy and emission spectroscopy will now be described.

After cutting the sample collected on a glass fiber filter or polystyrene filter according to the procedure shown in Fig. 3,1. the author prepared the sample solution for atomic absorption spectroscopy or emission spectroscopy by the procedure shown in Fig. 3.9. In principle, after ashing treatment with a low temperature ashing apparatus, sample solutions were prepared for Cr and Mo analysis by a fusion operation, while sample solutions

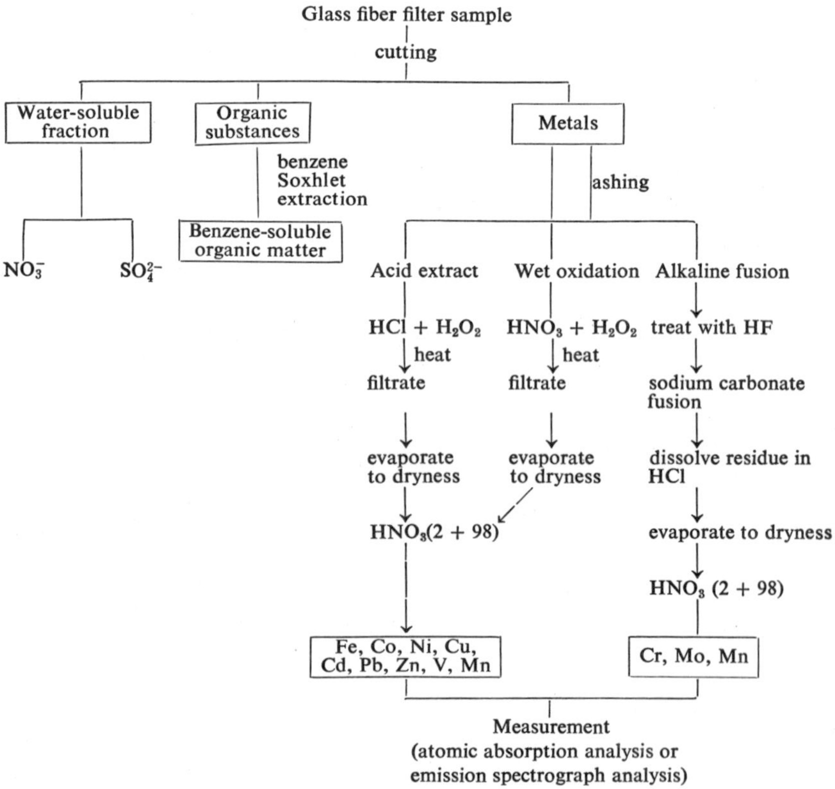

Fig. 3. 9. Flow chart for analysis of atmospheric particulate samples.

of other heavy metals were subjected to acid extraction. Samples with relatively little tarred organic substances and free carbon can be subjected to wet oxidative decomposition with HNO_3 and H_2O_2.

Ranweiler *et al.*[5], conducted an analysis of 22 elements including Si, Al, Ca, Fe, K, Na, Mg, Pb, Cu, Ti, Zn, Sr, Ni, V, Mn, Cr, Rb, Li, Bi, Ca Cs and Be by the procedure shown in Fig. 3.10. Sample collection was performed with a Microsorban filter (made of polystyrene) and a high volume sample was used. The recovery rate with this filter is shown in Table 3,7, and apart from Cd, Be and Al, extremely good recoveries and low standard

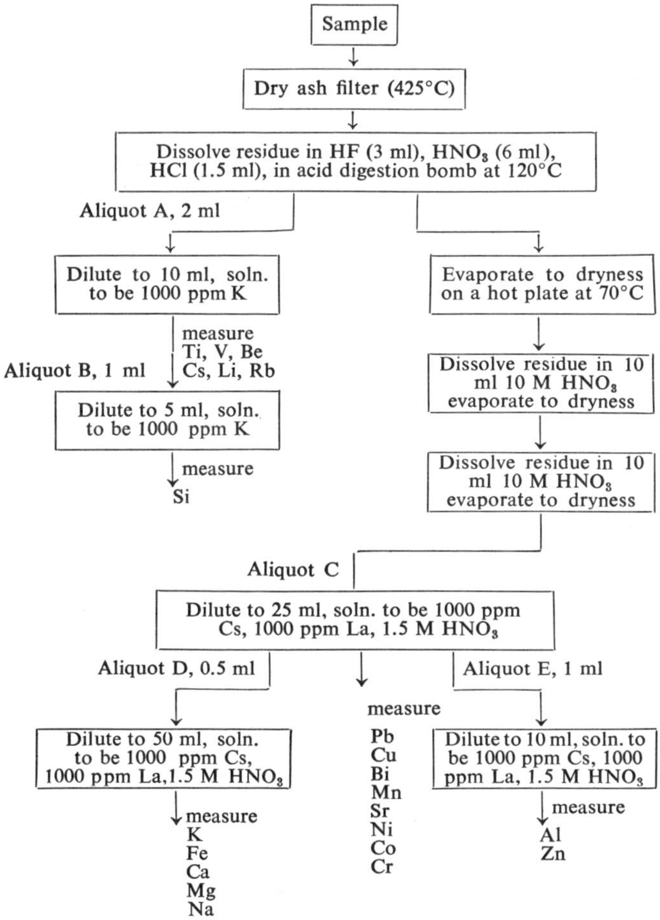

Fig. 3.10. Flow chart for atomic absorption analysis of atmospheric particulate samples. (Source: ref. 3. Reproduced by kind permission of the American Chemical Society, U.S.A.)

TABLE 3.7. Results of recovery tests using polystyrene filters

Element	Quantity supplied (μg)	Quantity recovered (μg)	Recovery rate (%)	Standard deviation (%)
Si	10,000	9,700	87.0	4.2
Al	6,108	5,815	95.2	1.2
Cu	2,000	2,038	101.9	4.3
Fe	3,000	2,856	95.2	3.4
K	4,000	3,972	97.3	2.9
Na	1,000	1,032	103.2	5.6
Mg	1,000	985	98.5	1.8
Pb	500	491	98.1	3.3
Cu	100	100	100.0	2.4
Ti	400	395	98.8	5.7
Zn	100	128	128.0	12.0
Cd	10	9.3	95.0	2.4
Sr	40	40.4	101.1	3.5
Ni	10	10.3	103.0	4.5
V	100	106.1	106.1	3.3
Mn	45	44.9	99.8	1.4
Cr	5	5.0	100.0	7.6
Co	5	5.01	100.2	16.9
Be	5	4.72	94.3	6.4

(Source: ref. 5. Reproduced by kind permision of the American Chemical Society, U.S.A.)

deviations were acquired with all elements. Moreover, the limits of quantitative analysis for each element in the case of a collected 2000 m^3 sample over a period of 24 h were also given (Table 3.8).

3.2. Preparation for analysis

A. Selection, preparation and storage of reagents and chemicals

a. Selection of reagents

In order to measure the concentration of metals in a given sample accurately, it is necessary first to prepare standard solutions from reagents of the highest purity. Problems generally do not arise when NBS or JIS (Japan Industrial Standard) special grade reagents are used; however, since elemental impurities are not always stipulated in writing, care is necessary.

When purchasing pure metals, it is important that the buyer know whether the reagent is 4 nine (e.g. 99.9954%) or 5 nine (e.g. 99.9995%), if the purity composition has been adequately analyzed. He should select materials

TABLE 3.8. Practical detection limits in the atomic absorption analysis of atmospheric particulate samples collected on polystyrene filters

Element	Limit ($\mu g/m^3$)	Flame
Si	0.8	N_2O–C_2H_2
Al	0.4	N_2O–C_2H_2
Ca	0.03	N_2O–C_2H_2
Fe	0.07	Air-C_2H_2
K	0.04	Air-C_2H_2
Na	0.03	Air-C_2H_2
Mg	0.01	N_2O–C_2H_2
Pb	0.007	Air-C_2H_2
Cu	0.003	Air–C_2H_2
Ti	0.07	Air–C_2H_2
Zn	0.03	Air–C_2H_2
Cd	0.0003	Air–C_2H_2
Sr	0.003	N_2O–C_2H_2
Ni	0.002	Air–C_2H_2
V	0.006	N_2O–C_2H_2
Mn	0.0003	Air–C_2H_2
Cr	0.0003	Air–C_2H_2
Rb	0.0005	Air–C_2H_2
Li	0.0003	N_2O–C_2H_2
Bi	0.002	Air–C_2H_2
Co	0.0008	Air–C_2H_2
Cs	0.00001	Air–C_2H_2
Be	0.00001	N_2O–C_2H_2

(Source: ref. 5. Reproduced by kind permision of the American Chemical Society, U.S.A.)

which bear a recognized certification of purity. Also, care should be given to the type of container, whether or not nitrogen has been used to displace remanant air prior to sealing, any bottle defects, and the utilization of a wrapping vinyl bag. In this respect, reagents bearing the JM standard (Johnson Matthey Chemicals Ltd.) have a high degree of reliability, since they carry a dated certificate of elemental impurity composition based on emission spectrographic analysis.

Pure metals come in a multitude of forms and only those which are of high purity and easily weighed should be selected for analytical use. Purchasable forms include powders, wire, sponge, foil, rods, ingots, chips, and crystal bar. In general, metals in the form of powder, sponge or wire are easily weighed. However, the former two are of slightly lower purity than the others.

Next comes the selection of the inorganic acids required for preparing the standard metal solutions and for decomposing, dissolving and extracting the samples. The most frequently used compounds include hydrochloric acid, nitric acid, sulfuric acid, hydrogen peroxide and perchloric acid. When samples are subjected to concentration procedures following solution and extraction, metal impurities in the inorganic acids are similarly concentrated, as mentioned above. This gives rise to important errors in experiments demanding a high degree of precision. In particular, sulfuric and perchloric acid have geen found to contain relatively high quantities of metal impurities. This may be attributed to their higher boiling points and viscosities relative to other inorganic acids and to the fact that their purification is extremely difficult. The specific gravity of commercial sulfuric acid is 1.84; the 96% acid has a boilding point of 280°C, while that for the 98.3% acid is 330°C.

The author has carried out decomposition treatment of atmospheric dust (suspended matter) by the wet oxidative decomposition method using sulfuric and nitric acid, and in the course of cadmium measurements by atomic absorption spectrophotometry of the extracted solution, large quantities of both Cd, and Ni were detected in blanks. Results of a careful check on the source of this contamination revealed that both metals occurred as impurities in the sulfuric acid. On the basis of a subsequent reexamination of special grade sulfuric acid manufactured by companies "W" and "J", Cd, Cu, Pb and Ni were detected as metal impurities in both cases. The results of another investigation of the metal impurities in sulfuric, nitric and hydrochloric acid are shown in Table 3.9. It should be noted that the concentrations of metal impurities in sulfuric acid were found to be especially high. It is thus imperative that blanks be run whenever sulfuric acid is used in analytical work, and in cases where alternative acids are adequate or suffice, sulfuric acid should be avoided entirely. Mercury has also been found as an impurity in sulfuric acid, a condition which all researchers or analysts should bear in mind when this acid is employed in mercury analysis.

Hydrochloric acid and nitric acid are relatively easily purified. However, quartz glass distillation apparatus should be used for the purpose

TABLE 3.9. Concentration of metal impurities (ppm) in inorganic acids

	Cu	Pb	Zn	Cd
Sulfuric acid	0.023	0.025	0.135	tr
Nitric acid	0.002	0.000	0.022	0.000
Hydrochloric acid	0.004	tr	0.027	0.000

since ordinary glass still invites the possibility of contamination. All glass instruments should be thoroughly washed and those with even the slightest defect should be discarded.

When purchasing inorganic acids, colored containers are undesirable since the possibility of iron contamination, among other impurities, exists. Hydrogen peroxide for analytical work should be purchased in polyethylene bottles.

b. Preparation of standard solutions

Preparative methods for the standard solutions employed in metal analyses are listed in Table 3.10 (final concentrations, 1 mg/ml). Since the prepared concentrations of given standard solutions cannot be reliably preserved for long periods, fresh solutions should be prepared for successive analyses.

c. Reagents, standard solutions and preservation of sample solutions

In cases where pure metals are used as reagents, the air in the container should be displaced immediately with very pure nitrogen gas so that oxidation is precluded, following removal of the required quantity. For storing standard solutions, polyethylene containers have been found the most suitable. One shortcoming of glass containers is the occurrence of alkaline dissolution. Such containers should therefore be used with caution or avoided entirely if possible, although problems generally do not arise with Pyrex #7740. Even when standard solutions are stored in polyethylene containers over long periods, the titanium and aluminum which are employed as catalysts in the manufacturing process tend to leach out the inner wall of the container. Thus, although polyethylene containers have been generally regarded as the best all-round receptacles for the storage of reagents and standard solutions, careful consideration of the merits and demerits of various vessels should be given in selecting the most suitable one for the intended purpose.

Concerning the storage of sample solutions, both polyethylene and Pyrex test tubes serve well. Such test tubes should be thoroughly washed in nitric acid prior to use in order to avoid alkaline dissolution from the wall. However, long-term storage of sample solution extracts is itself undesirable, and samples should be submitted to analysis as soon as possible.

Particular attention should be given to the following during the storage of sample solutions.

(1) Employment of containers that are tolerant to acid attack.

(2) Meticulous and thorough washing and drying of containers and

TABLE 3.10. Methods for the preparation of elemental standard solutions (1 mg/ml)

Element	Standard substance	Preparative method for standard solution
Ag	$AgNO_3$	$AgNO_3$ is dried at 110°C and 1.575 g weighed out and dissolved in 1 l water (final vol)
Al	Metallic aluminum	Metallic Al (1.000 g) is dissolved by heating in a slight excess of diluted HCl and made up to 1 l after cooling
	Alum	1.757 g of $Kal(SO_4)_2$. 12 H_2O is dissolved in 100 ml of 0.25 N HCl
As	Arsenic trioxide	1.321 g of As_2O_3 is dissolved in 2 ml of 1 N NaOH. After diluting in water, it is slightly acidified with HCl and then made up to 1 l with water
Ba	Barium chloride	1.779 g of $BaCl_2$. $2H_2O$ is dissolved in 1 l of water (final vol)
Be	Metallic beryllium	1 g of metallic Be is dissolved by heating in 10% HCl and made up to 1 l after cooling
Ca	Calcium carbonate	2.497 g of $CaCO_3$ is dissolved in an excess of diluted HCl and made up to 1 l with water
Cd	Metallic cadmium	1.00 g of metallic Cd is dissolved in diluted HNO_3 and the solution evaporated over a water bath; †[1] 5 ml of 1 N HCl are added and the mixture reevaporated over a water bath. 1 H NCl and water are added to dissolve the dried substance and the solution made up to 1 l with water
	Cadmium oxide	1.142 g CdO is dissolved in 25 ml of HNO_3 and made up to 1 l with water
Cr	Metallic chromium	1.000 g of metallic Cr is dissolved in 2 N HCl and made up to 1 l with water
Cu	Metallic copper	1.000 g of copper wire or flakes is dissolved in 25 ml of 40% HNO_3, boiled and after removal of nitrogen oxides, the remainder is diluted with water to make 1 l
Fe	Metallic iron	1.000 g of iron wire is dissolved in 25 ml of 40% HNO_3 and after nitrogen oxides have been expelled, is diluted in water to make 1 l
K	Potassium chloride	1.907 g of KCl is dissolved in water to make 1 l
Mg	Metallic magnesium	1.000 g of metallic Mg ribbon is dissolved in 100 ml of dilute HCl (1:100)†[2] and water is added to make 1 l
Mn	Metallic manganese	1.000 g of metallic Mn is dissolved in 10 ml of 40% HNO_3 and subjected to evaporative drying over a water bath, after which 5 ml of HCL are added followed by a second evaporative drying. A small amount of HCl is added to redissolve and the solution made up to 1 l with water

TABLE 3.10——*Continued.*

Element	Standard substance	Preparative method for standard solution
Mo	Metallic Mo	1 g of metallic Mo is dissolved in a small amount of aqua regia, subjected to evaporative drying over a water bath, and then dissolved in HCl and made up to 1 l with water
	Mo oxide	MoO_3 (1.50 g) is dissolved in a small amount of ammonia water and made up to 1 l with water
Na	Sodium chloride	2.541 g of NaCl is dissolved in water to make 1 l
Ni	Metallic nickel	1.000 g of metallic Ni is dissolved in dilute HNO_3, subjected to evaporative drying over a water bath, dissolved in HCl, and then made up to 1 l with water
	$NiCl_2.6H_2O$	0.405 g of $NiCl_2.\ 6H_2O$ is dissolved in 1 l of 0.1 N HCl (0.1 mg Ni/ml)
Pb	Lead nitrate	1.598 g of $Pb(NO_3)_2$ is dissolved in HNO_3 (1:100) and made up to 1 l with water
Pd	Metallic palladium	1.000 g of metallic Pd is dissolved in a small amount of aqua regia, evaporatively dried over a water bath, dissolved in 1 N HCl, reevaporatively dried, dissolved in 1 N HCl, and made up to 1 l with water
Sb	Antimonyl potassium tartrate	2.6684 g of $K(SbO)C_4H_4O_6.\ 1/2H_2O$ is dried in an oven at 105°C for 2 hr, dessicated to constant weight, and then dissolved in water to make 1 l
Se	Selenium dioxide	1.405 g of SeO_2 is dissolved in water to make 1 l
Si	Silicon dioxide	0.2141 g of SiO_2 is fused with 2 g Na_2CO_3 in a platinum crucible and dissolved in water to make 100 ml
Sn	Metallic tin	1.000 g of metallic Sn is dissolved in 6 N HCl and diluted to 1 l with water. In certain instances, it is recommended that thioglycol be added as a reducing agent†3
V	Vanadium pentaoxide	1.7849 g of V_2O_5 is dissolved in dilute HCl and made up to 1 l with water
Zn	Metallic zinc	1.000 g of metallic zinc is dissolved in a slight excess of dilute HCl and made up to 1 l with water
	Zinc oxide	1.245 g of ZnO is dissolved in 10 ml of 40,% HNO_3, dissolved in 25 ml water, made up to 1 l with water

†1The melting point of $Cd(NO_3)_2 4H_2O$ is 59.4°C, and boiling point 132°C; thus, heating to 100°C should not be done on a hot plate. A water bath at a temperature of 80–90°C is suitable.

†2When concentrated HCl or a large quantity of HCl is added all at once, the reaction is violent, bumping and spillage is likely to occur and careful attention is required.

†3Long-term storage of tin of valence 2 is difficult. When thioglycol is added, storage can be continued for 2 weeks. Material used in calibration is good for only 1–2 days, so that new standard solutions for use in calibration must be prepared daily.

instruments prior to use; acid washing (soaking in 20% nitric acid solution) of all new glass instruments and receptacles.

(3) Ensurance of proper entry of the classification code for each sample with details written on affixed labels.

B. Water purification

Care should always be taken to ensure water purity in quantitative analyses requiring the utmost precision. Purified water is obtained by distillation or passage through an ion exchange resin. In the purification process, it is recommended that a continuous distillation apparatus made of either quartz or borosilicate glass, a copper distillation apparatus, or an ion exchange apparatus be used.

A copper still is more efficient than one made of glass, but considerable copper and zinc solution (contamination) occurs in this case, so that the water distilled cannot be used for the analysis of metals. In some cases, the inner surface of the copper still has a coating of tin, but when this is damaged, various kinds of heavy metals escape and dissolve in the water.

The most frequently employed ion exchange method involves the use of a pure water ion exchange apparatus with which pure water can be prepared by passing tap water through an equivalent mixture of both strongly acidic and strongly basic cation exchange resins. Under conditions where extremely pure water is required, it is necessary to distill ordinary water at least twice or to distill once with a borosilicate glass distilling apparatus, and then to pass the distilled water through an ion exchange apparatus. There are also methods where the reverse procedure is used, i.e. passage through an ion exchange system first, then distillation; however, in this case, the effective life of the resin is shortened. The electrical conductivity of distilled water is some 2–8 $\mu℧$/cm 25°C but becomes 0.7 $\mu℧$/cm 25°C after ion exchange treatment. Highly pure water acquired in this manner readily becomes contaminated when introduced into containers that are poorly cleaned or of inferior quality. Glass containers should be avoided since dissolution of alkaline materials is always possible.

C. Glass containers

The most important single property of the glass used in chemical analysis is a high resistance or tolerance to various chemicals, i.e. water, acids, bases, salts, etc. Also, the heat resistance, facility in processing and transparency should be carefully considered when selecting an appropriate type.

a. Classification of glass according to chemical composition

The types of glass used in chemical research may be classified as follows.

Soda lime glass (common glass): this is frequently used in the manufacture of test tubes, flasks, beakers, glass tubes, etc. Its molar composition is Na_2O, CaO, $6SiO_2$, and it is characterized by a low corrosion resistance, heat resistance up to *ca.* 400°C, and high alkaline dissolution.

Potash lime glass: this has a slightly higher melting point than soda lime glass and is characterized by a high corrosion resistance. Its molar composition is K_2O, CaO, $6SiO_2$. Moreover, glass of this type containing equivalent amounts of Na_2O and K_2O is also available, and has excellent chemical durability.

Borosilicate glass (hard glass): this is generally called "hard glass' and has a molar composition of Na_2O, CaO, Al_2O_3, B_2O_3, $6SiO_2$. Its melting point is high and it has superior heat resistance and durability. Included in this type of glass is the well known product, "Pyrex". In Pyrex #7740, the SiO_2 content is extremely high; the B_2O_2 content is also high, and due to the low content of Na_2O and K_2O, the coefficient of linear expansion is very low. Borosilicate glass is also excellent in terms of heat resistance and has a melting point of 820°C (*cf.* Table 3.11).

TABLE 3.11 Composition of borosilicate and other glasses (%)

Composition Product name	SiO_2	Al_2O_3	Fe_2O_3	B_2O_3	Na_2O	K_2O	ZrO_2	As_2O_3
Pyrex #7740	80.8	2.3	0.03	12.5	4.0	0.4	0	0
Vycor #7913	96.7	0.24	0	2.65	0.02		0.4	0
Terex (Japan)	80.8	2.3	0.03	12.5	4.0	0.4	0	0.07

Vycor glass (high silicate glass): this possesses superior heat resistance and chemical durability similar to quartz glass. It contains 96–98% SiO_2 as well as 5% B_2O_3 and 0.5% Na_2O.

Phosphate glass: P_2O_5 is the primary constituent but it also contains Al_2O_3, B_2O_3, ZnO, MgO and PbO. This type of glass is characterized by a high durability to hydrofluoric acid and is employed in experiments where this acid is used.

Quartz glass: this is made from quartz or rock crystal by melting at a temperature above 2000°C. It consists essentially of SiO_2 but does contain detectable amounts of Al_2O_3, Fe_2O_3, B, Na_2O, K_2O, CaO, TiO_2, Sb, As, Co, Cr, Mn, P, Zr and Ga. In particular, the rock crystal used in standard grade quartz glass is of inferior quality, but no problems arise with the highest grade material. Quartz is resistant to acid, but caution is required since it is not tolerant to alkali.

In addition to these representative types of glass, numerous other types exist such as lead alkaline glass, silicate glass, alumina silicate glass, etc.

b. Selection of glass

The most important single consideration in metals analysis is the selection of glass instruments which are highly resistant to corrosion. The amount of free silicate in glass having such characteristics is high and the corrosive resistance is greater than in glass having a lower content of alkali metals oxides or silicates. In general, glass is easily damaged by hydrofluoric acid and hot concentrated alkali. Moreover, it is more easily damaged by weak than concentrated acid, since the silicates of alkali metals readily dissolve in aqueous solutions of dilute acids. In addition to the glass compositions given above (*cf.* Table 3.11), AsH_3 and various metallic compounds are added in some cases to the fusion pots in order to facilitate the removal of gas bubbles and to increase transparency. Furthermore, since the fusion pots are usually small, the amount added varies according to and is determined by the appearance of the mixture as it forms. Careful consideration must therefore be given to the selection of glasses manufactured by even the same company, since the amounts of such chemicals vary according to lot. However, chemicals for gas bubble removal are not added to Pyrex glass, and since this type of glass is mass produced, its quality is essentially uniform.

D. Contamination arising from containers, instruments and washing

Careful attention must be paid to reagents, containers, instruments, laboratory conditions, and even to the researcher himself, since all are possible sources of contamination during the analysis of trace quantities of metals.

a. Contamination arising from poorly cleaned containers and instruments

All glass containers and instruments should be immersed in acid solution before use, to elute alkaline substances. When glass containers are employed, sample contamination and loss occur to some extent due to ion exchange at the inner surface of the container. If scratching or corrosion occurs on the container after use, contamination and loss are especially augmented as a result of solution of metallic constituents, surface adsorption and ion exchange, a situation which inevitably leads to variations in data and spurious analytical results. All glass appliances used in acid extraction, etc. should therefore be discarded after several weeks of use.

Surface adsorption occurs to a greater extent with teflon and polyethylene than with other materials. Containers and instruments manufactured from these synthetics are therefore easily contaminated. When teflon containers are washed, dried and allowed to stand for some time, the surface becomes darkened. Containers and instruments having such properties should be kept in dry storage receptacles (*cf*. Fig. 3.11, below).

Table 3.12 summarizes results of a neutron activation analysis of contamination (metal impurities) derived from containers, reagents and filters used in the quantitation of trace metals in sea water by Robertson.[6)] Particular attention should be given to employment of the following materials in analytical work.

Teflon: it is noteworthy that essentially no metal constituents elute from this material, except for some Na; moreover, it can be used over a wide temperature range, −200 to 204°C.

Polyvinyl chloride: since several metals such as Zn, Fe, Sb, Co, etc. are detectable, very careful use of this material must be made when these metals are analyzed.

Nylon: Co has been detected in slightly greater quantities than in other materials; this is attributed to the use of Co compounds in the manufacturing process.

Quartz glass containers: metal elution occurs to a lesser extent than with hard glass; furthermore, the metal content of synthetic quartz glass is lower than that of natural quartz glass.

Vycor tubing: the metal content is extremely low.

Polyethylene bottles: these are considered excellent since elution of the metallic fraction is practically nil. However, there is relatively greater surface adsorption than with other materials; this undesirable characteristic can be suppressed somewhat by immersion in hydrochloric acid (pH 1.5) prior to use.

Millipore filters: Zn, Cr, etc. have been detected. Care should therefore be exercised, although this is not a serious factor in the analysis of atmospheric dust.

Kim wipe tissue paper: this is frequently used to blot water from flasks and pipettes and particular attention should be given to Zn contamination. Products free from lint or which leave no lint after use are now available for experimental work.

Redistilled water: very little difference exists between distilled water prepared with a quartz still and redistilled water prepared by other means.

Nitric acid and hydrochloric acid: the purity is usually high but can be improved by redistillation. Moreover, when redistillation is performed, it is recommended that a slightly used rather than new distillation apparatus be employed since contamination from the apparatus is thereby minimized.

TABLE 3.12. Content of metal impurities in reagents, materials and containers

	Zn	Fe	Sb	Co	Cr	Sc	Cs	Ag	Cu
Materials and containers									
Teflon	9.3	35	0.4	1.7	<30	<0.004	<0.01	<0.3	22
Polyvinyl chloride	7120	270,000	2690	45	2	4.5	<1	<5	630
Polyethylene tube	55	7.4	9000	140	254	11	<100	<300	N.M.†[1]
Quartz tube	1–33	395	0.01–1940	0.44–12	2.5–6.02	0.03–0.39	0.1–1390	0.01–0.1	0.03–2.0
Borosilicate glass	730	280,000	2900	81	U.M.	106	<100	<0.001	N.M.
Vycor glass	U.M.†[2]	U.M.	1.09×10^6	U.M.	U.M.	U.M.	U.M.	U.M.	U.M.
Polyethylene	28	10,400	0.18	0.07	76	0.008	<0.05	<0.1	6.6
Kim wipe tissue paper	48,800	1000	16	24	500	14	<0.1	~0.8	N.M.
Millipore filter	2370	330	39	13	17,600	0.79	1.5	<0.05	N.M.
Reagents									
Distilled water (quartz)	~1	~1	~0.06	~0.04	~2	0.0022	<0.01	<0.02	N.M.
Redistilled water	~1	<0.2	<0.01	<0.02	~2	<0.0001	<0.01	<0.02	N.M.
Redistilled water (3X)	0.5	~1	~0.02	<0.02	12	~0.0002	<0.01	<0.02	N.M.
Nitric acid	~13	~2	~0.03	0.018	72	0.0007	<0.01	~0.24	1.3
Redistilled nitric acid	~2	~1	~0.04	~0.03	13	~0.008	<0.1	0.29	N.M.
Hydrochloric acid	22	~1	0.20	0.09	1.1	0.002	<0.002	<0.1	82
Redistilled hydrochloric acid	~1	~1	~0.04	~0.08	6	~0.001	<0.01	<0.02	N.M.
Ammonia water	2.3	<0.1	<0.006	~0.009	<0.04	<0.0003	<0.002	<0.1	6.0
Dithizon	1150	<7000	0.8	1.2	<2000	0.15	10	<10	420
Cupferron	7000	<600	0.3	0.68	<200	0.04	<1	~3	160
DDTC–Na	40	<600	42	0.56	U.M.	0.02	<1	<10	N.M.
Sodium hydroxide	<40	<900	<0.32	5.5	60	0.30	0.69	<0.2	N.M.
Potassium hydroxide	1250	2700	1.8	1.7	<10	0.04	<0.01	66	N.M.

†[1] N.M. = not measured.
†[2] U.M. = unmeasurable.
(Source: ref. 6. Reproduced by kind permission of the American Chemical Society, U.S.A.)

Dithizon, Cupferron: chelates of these compounds contain relatively large amounts of Zn, Co, Fe and Ag impurities; however, the amounts used for analysis are minute, so that the contamination is not critical.

b. Washing of apparatus and instruments

A mixture of potassium dichromate and sulfuric acid has found wide application in laboratories as a cleaning solution for glass instruments. It is especially suitable for the removal of organic materials adhering to the instrument surface. However, in the analysis of trace metals, even after washing the instruments, Cr is found to remain on the surface. This cleaning solution should therefore be avoided.

In the analysis of trace metals, the following cleaning solutions are recommended.

Sulfuric acid-nitric acid mixture: concentrated sulfuric acid and concentrated nitric acid are mixed in a ratio of 1:1. The resultant solution possesses a cleaning power comparable to potassium dichromate-sulfuric acid solution.

20% nitric acid solution: generally speaking, most adherent grime is effectively removed by this cleaning solution, but if followed by cleaning with 35% hydrochloric acid solution, all persistent grime is also completely removed. These acids can be kept and stored in polyvinyl chloride tanks designated for the storage of sulfuric acid (diameter 30 cm, height 45 cm).

Apart from the above-mentioned cleaning methods, other procedures employ high frequency sound waves (sonification). Also, for contamina-

TABLE 3.13. Cleaning methods for the removal of various pollutant materials from instruments and apparatus

Substance	Cleaning solution
Fats	Carbon tetrachloride or similar organic solvents
Organic substances	Warm concentrated detergents; hot concentrated sulfuric acid containing a few drops of sodium and potassium nitrite
Oxides of copper and iron	Potassium chlorate
Mercury sulfide	Hot aqua regia
Silver chloride	Ammonia; thiosulfate
Aluminum residual silicates	Cleaning in 2% hydrofluoric acid followed by cleaning in concentrated sulfuric acid, immediately after which the instrument is well rinsed with distilled water; next, washing with a few ml of acetone; the procedure is repeated until residues are thoroughly removed
Barium sulfate	Concentrated sulfuric acid at 100°C
Residual mercury compounds	Hot concentrated sulfuric acid
Stannous sulfate	Boiled sulfuric acid

tion by certain characteristic substances, various cleaning methods are used (see Table 3.13).

E. Contamination by laboratory dust and its prevention

Washed instruments should be handled in such a way that contamination is prevented, and dried with similar care. After cleaning, glassware should be rinsed well in distilled water to remove all traces of cleaning solution and placed on a drying rack with a gauze covering to prevent contamination. Immediately after drying, it is recommended that the openings of flasks, test tubes and pipettes be wrapped in pharmaceutical paper, then stored in a dry receptacle (40 × 40 × 40 cm, Fig. 3.11). In particular, instruments made of polyethylene and teflon which are characterized by high surface adsorption, should always be stored under dry conditions.

a. Prevention of contamination during the concentration procedure

When extracts or dissolved solutions are subjected to a concentrating procedure, the operation should be carried out in a well ventilated draft

Fig. 3.11. Rectangular type plastic drying rack.

chamber using a hot plate or water bath. However, since such concentration often requires considerable time, the possibility of occurrence of contamination is great. The following method is considered suitable for concentrating extracts or dissolved solutions in the shortest possible time and with the minimum of contamination.

The apparatus is shown in Fig. 3.12. A flat, circular, hard glass plate is placed on a hot plate and a glass cylinder (diameter 25 cm, height 16 cm) is positioned on top of it. After placing the beaker which contains the sample in the cylinder, a cover is positioned on top of the cylinder and clean compressed air is delivered into the chamber via an inlet near its base.

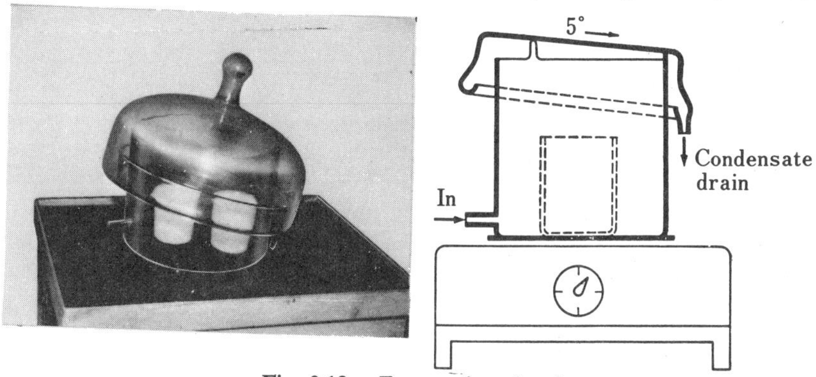

Fig. 3.12. Evaporation chamber.

By methods such as this, contamination can be kept to a minimum and the operation is itself rapid. However, when the number of samples to be treated is large, such methods are unsuitable or impractical. This difficulty can be overcome with procedures such as concentration by vacuum lyophilization, which is based on the principle used for producing freeze-dried coffee. More specifically, a freezing mixture of dry ice and acetone is put into a cold trap, and a flask containing the sample frozen onto the inner wall is connected to the arm of the cold trap. When the pressure is reduced via the exhaust pipe, water (moisture) rapidly sublimates from the surface of the frozen sample which becomes lyophilized. This freeze drying method has been employed in the fields of biochemistry and food technology and should find wide application in the field of atmospheric pollution since 4–6 samples can be processed at one time and the treatment can be done rapidly, without contamination. However, it is unsuitable for condensation of acidic extracts.

b. Laboratory conditions and draft chambers

A few years ago, the author surprisingly found that draft chambers previously thought to be essentially free of contamination, were in fact more contaminated than the laboratory room. This was apparent from an examination of the degree of contamination arising from dust present in both a laboratory room and draft chamber in the case of polyethylene vats containing 2 N hydrochloric acid which had been left standing for several days.

It is often the case that insufficient attention is given to contamination arising from laboratory room and draft chamber conditions. A dust-free room (clean room) is considered ideal for trace analysis; however, to build such a laboratory, enormous funds are required and construction is difficult. As a reasonable alternative, therefore, laboratories generally operate on a system whereby air is passed through a filter and the floor dust is vacuumed rather than swept.

Water baths should be used under a hood, and exhaust gas should be made to flow in an upward direction by means of an ascending vapor flow and hood fan. However, since the ascending vapor flow and amount of exhaust gas from the bath are not in balance, gas and water vapor sometimes escape from the hood to the outside. In this case, special care must be exercised.

In the preparation of electrodes and samples used for emission spectral analysis, a "clean bench" (dustless work bench) of the type found in semiconductor, watch and camera factories is recommended.

The interior of draft chambers has generally been thought to be extremely clean and has been designed primarily for the exhaust of poisonous gases without consideration of the elimination of laboratory dust. In reality, however, draft chambers are generally contaminated by dust to a greater degree than the laboratories themselves. Moreover, opening and closing of the draft chamber window creates disturbances in the air currents within the chamber, so leading to further risk of sample contamination. For example, when a researcher stands in front of the draft chamber window, his presence creates current disturbances, and even if the window is closed, the position of the exhause port and its degree of aperture or closure determine the extent to which disturbances of air currents arise. However, if an air curtain arrangement is employed in conjunction with the type of draft chambers currently used, disturbances in the air flow can be minimized.

References

1. E. I. Hamilton *et al.*, *Analyst*, **92**, 257 (1967).
2. T. Y. Kometani and J. L. Dove, *Environ. Sci. Technol.*, **6**, 617 (1972).
3. C. E. Gleit and W. D. Holland, *Anal. Chem.*, **34**, 1454 (1962).
4. K. Oikawa *et al.*, *Bunseki Kagaku* (Japanese), **26**, 630 (1976).
5. L. E. Randweiler and J. L. Moyers, *Environ. Sci. Technol.*, **8**, 152 (1974).
6. D. E. Robertson, *Anal. Chem.*, **40**, 1067 (1968).

CHAPTER 4

ANALYTICAL METHODS FOR METAL COMPONENTS

4.1. Principles of metal analysis

Analyses of the metal content of environmental pollutants are performed by various methods such as atomic absorption spectrometry, emission spectroscopic analysis, X-ray analysis, neutron activation analysis, etc. Taking into account the present status and probable future development into account the present status and probable future development of instrumental analysis, a possible scheme for metal analysis is shown in the form of a flow sheet in Fig. 4.1.

In the case of analyzing environmental pollutants, the elemental analysis should not be restricted simply to detecting some expected elements, but should preferably cover multiple elements from the same material also. This can be done by X-ray fluorescence analysis, X-ray microanalysis and emission spectroscopic analysis. The co-existing pollutant elements (matrix) can then be assessed, which is valuable knowledge in the subsequent analytical procedures. Use of X-ray fluorometry or an X-ray microanalyzer permits non-destructive analysis, so that the analyzed materials can be kept as a specimen or in their natural form. In the case of emission spectroscopic analysis, the photo-plate can be kept as a specimen. In environmental pollutant analysis, the obtained data are of course of paramount importance, but leaving the pollutant as a raw material is also important.

When qualitative or estimation analysis has been made, quantitative analysis should follow. The most widely used method for the analysis of metals is atomic absorption analysis. Recent developments in the analytical instrumentation have greatly enhanced the sensitivity and accuracy, so that almost all environmental pollutants can now be analyzed.

Emission spectrochemical analysis allows simultaneous multiple element analysis, both qualitatively and quantitatively. The difficulty is that it requires sophisticated techniques, although the sensitivity and accuracy increase greatly when a plasma-jet or photomultiplier apparatus is equipped. Considering non-destructive analysis, X-ray fluorescence analysis is not always sufficiently sensitive, but elements greater than atomic number

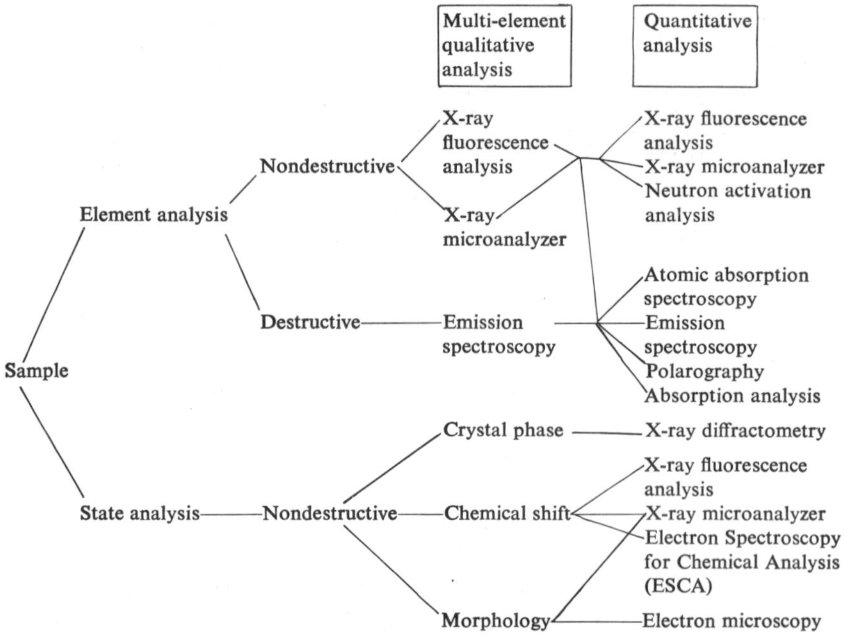

Fig. 4.1. Proposed flow sheet for metal element analysis.

9 (fluorine) are analyzable. In the case of atomic absorption spectrometry and emission spectrochemical analysis, the samples should be solubilized in liquid form. Elements such as Cr, Ti, Si, and Al should thus be prepared by alkali fusion, but this is quite difficult. Such elements are better analyzed by X-ray fluorescence analysis. Neutron activation analysis is a nondestructive technique that is also highly sensitive. The semiconductor detector (Ge (Li) detector) permits high resolving power and has led to great progress in simultaneous multiple element analysis. Recently, a computerized automatic analyzing system was established.

Many reports have been made on the results of metal component analyses of atmospheric particles. However, these have tended to deal with the metal component essentially as "metal", i.e. the metal as an element. It is important also to clarify the actual state of the metals in the atmosphere, so that their effects on the biological environment, especially on man, can be considered more precisely. The physical and chemical properties of atmospheric particles, and the origin of environmental pollutants, should also be made clear. However, very few reports yet exist on the state of atmospheric pollutants.

X-ray diffractometry permits quantitative analysis in the crystal phase. The chemical shift of many elements can be confirmed by X-ray fluorescence analysis or with an X-ray microanalyzer.

Generally speaking, the analytical methods of the future will be highly automated and multiple elements will be detected simultaneously within a shorter period. The samples should be prepared non-destructively over a shorter period, and human errors should be eliminated by the exclusion of handling.

4.2. Homogeneous dust sample (AS-1) for comparison of the accuracy of analytical results on atmospheric samples

Many reports have described metal contamination of the atmosphere, using atomic absorption spectrometry, emission spectrochemical analysis, X-ray fluorescense analysis, neutron activation analysis, and mass-spectrography. However, until the accuracy of the analytical results can be established, the data remain mere numbers and cannot be regarded as reliable

TABLE 4.1. Analytical results of aliquots of series A by atomic absorption spectrometry (units, μg/g)

No. of aliquot	Element			
	Zn ($\times 10^3$)	Cu ($\times 10^2$)	Pb ($\times 10^3$)	Mn ($\times 10^3$)
x_1	3.13	3.5	2.6	1.07
x_2	3.11	3.4	2.6	1.06
x_3	3.10	3.3	2.7	1.09
x_4	3.05	3.4	2.6	1.08
x_5	3.09	3.6	2.6	1.10
x_6	3.03	3.4	2.6	1.08
x_7	2.96	3.3	2.6	1.08
x_8	2.97	3.5	2.8	1.06
x_9	2.98	3.5	2.6	1.10
x_{10}	3.00	3.4	2.7	1.09
x	3.04	3.4	2.6	1.08
$V_{\text{anal+mix}}$†	3970	90	4900	210
CV(%)	2.1	2.8	2.6	1.3

† $V = \sum_{i=1}^{n}(x_i - \bar{x})^2/(n-1)$.

(Source: ref. 5. Reproduced by kind permission of the American Chemical Society, U.S.A.)

TABLE 4.2. Results of interlaboratory analysis of AS-1 (units, $\mu g/g$)

	Analytical method†[1] and no. of detn.						
	INAA†[2]	INAA†[3]	NAA†[4]	INAA†[5]	XF†[6]	AAS†[5]	AAS†[7]
Element	6	10	4	5	1	10	5
Ag	3	3.7					
Al	38000	55000	56000	49000			
As	60	43		27			
Au	0.09		0.42	0.4			
Ba		680		140			
Br	350	350		330			
Ca		63000	56000	50000			
Cd						19	19
Ce		33		26			
Cl		33000	34000	26000			
Co	23	30	30	22			
Cr	370	360	330	360	300		
Cs	3.8	4.1					
Cu		600		320	2100	340	320
Eu		0.79		0.77			
Fe	47000	48000	39000	45000	47000	43500	44400
Ga				13			
Hf	3.3	3.4					
Hg	9						
K		9700		9600			
La	21	17		16			
Lu		0.3					
Mg		15000	16000	19000			
Mn	1200	1300	1300	1300	900	1080	
Na	12000	15000	13000	14000			
Ni		250		210	200	175	164
Pb					1800	2640	2030
Rb		45					
Sb	39	44	29	43			
Sc	12	8.6	12	10			
Se	2		16	10			
Sm	3.2	3.6					
Ta		1.2					
Th	3.9	6.2					
Ti		5800	4500	3700	2900		
V	200	290	240	240	240	100	300
W	3.2	28		33			
Zn	2800	3800	4100	3500		3040	

†[1] INAA, Instrumental neutron activation analysis; NAA, neutron activation analysis; XF, X-ray fluorescence analysis; AAS, atomic absorption spectrometry. †[2] Japan Atomic Energy Research Institute. †[3] Radiation Center of Osaka Prefecture. †[4] Atomic Energy Research Institute, Rikkyo University. †[5] Faculty of Engineering, Keio University. †[6] Kanagawa Prefecture Environmental Center. †[7] Japan Environmental Sanitation Center.
(Source: ref. 5. Reproduced by kind permission of the American Chemical Society, U.S.A.)

estimates. Use of a standarized sample is important for studying the analytical methods, for examining the accuracy of the data obtained by different analytical methods, and for comparing data between different analysts and/or different laboratories. In this connection, the "geological power sample" obtainable from the U.S. Geological Survey is well known and many studies have been performed systematically in various parts of the world on it. Results for G-1, W-1, G-2, PCC-1, and DTS-1 have been given.[1,2] Other standard samples such as orchard leaves[3] and bovine liver[4] can be obtained from the U.S.N.B.S., and many data have been accumulated by various researchers.

The author's group has prepared a standard sample from atmospheric aerosol and obtained good results.[5] First, a large amount of airborne particulate (6 kg) was collected from the entrance filter of an air-conditioning apparatus serving a building. The sample was dried, mixed thoroughly, and part was analyzed by atomic absorption analysis to obtain the quantities of Zn, Cu, Pb, Mn, etc. The results were examined by a statistical method to certify high homogeneity. Some of the data are shown in Table 4.1. This standard sample was named AS-1 (atmospheric sample No. 1), and is now analyzed by many workers and research institutes, not only by atomic absorption analysis but also by X-ray fluorescence analysis and neutron activation analysis. Results are gradually accumulating, and some are given in Table 4.2.

4.3. Atomic absorption spectroscopy

Atomic absorption spectroscopy was first introduced in 1955 by A. Walsh in Australia. Thereafter many studies were made on instrumental improvements, so that its field of application was greatly enlarged. Atomic absorption spectroscopy has now become a common analytical tool in many laboratories, due to its high sensitivity, accuracy and reproducibility, the ease of sample preparation and handling (incorporating less human error), and the comparatively low cost. Recently, environmental contamination by heavy metals has been given serious debate, and the introduction of atomic absorption spectroscopy was so welcomed by many researchers.

The principle of atomic absorption spectroscopy is first to vaporize atomically the metal-containing material by some appropriate method. The bottom state atom absorbs a specific wavelength of light transmitted through this atomic vapor layer. The absorption is analyzed quantitatively using a photomultiplier apparatus.

A. *Atomic absorption spectroscopy using a flame*

The traditional type of atomic absorption spectroscopy incorporates a furnace apparatus, as shown in Fig. 4.2. This consists of a nebulizer, disperser and burner head. The sample is introduced into the sample atomizing apparatus (burner), into the flame. Here, the expected element is ionized to yield free atoms. On the other hand, flameless methods are available. One is the cold vapor method usually used in mercury analysis, in which mercury vapor produced in the sample solution by reduction is passed into a gas cell with air and the absorbance measured. Another technique is to atomize the element using a graphite (heated) furnace. These flameless methods will be described in more detail later.

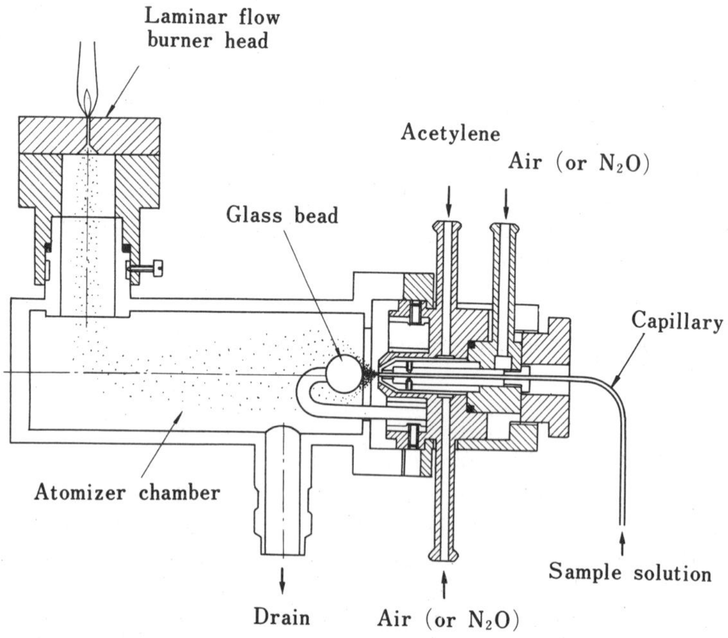

Fig. 4.2. Example of sample atomizer apparatus (flame method).

a. Accuracy and sensitivity of analysis

The analytical methodology of atomic absorption spectroscopy is comparatively easy and the human error is small. The excitation energy has no effect at all in contrast to emission spectrochemical analysis. The accuracy of analysis is very good since the bottom state, which has a steady electron configuration, is measured. However, there is variability in accuracy between different instruments and different elements. That of the single beam type depends on the stability of light emission from the lamp: the noise level is higher and the accuracy lower than in the double beam type. The criteria governing the accuracy are (1) stablility of the light source (hollow cathode lamp), which depends on an appropriate current, (2) smooth nebulization, and (3) cleanliness of the burner. The sensitivity is defined by the concentration of metal solution showing 1% absorbance and is expressed in μg/ml/1%. The factors affecting sensitivity are (1) the state of the flame, (2) selection of burning gas, (3) efficiency of the nebulizer, (4) working condition of the light source, (5) optical system and (6) dispersion ability of the diffraction grating. Many elements can be anlayzed to greater sensitivity than in emission spectroscopic analysis. However, there are differences according to excitation method and instrument, so that details cannot be compared summarily. The sensitivity for each element is shown in Table 4.3. Extraction from the sample solution by organic solvents permits an increase in sensitivity, as will be described later.

b. Interference

Interference in atomic absorption analysis can be explained optically, chemically and physically.

i) Optical interference

Optical interference is represented by molar absorption and optical dispersion. When NaCl or some other alkali metal halide is present in the sample solution at a high concentration (over 500 ppm), unsaturated compound or atomic group appears in the flame, bringing about molar absorption and optical dispersion in the frequency range of 200–300 nm. This results in a higher absorbance than the true level (Fig. 4.3). In such cases, the object metal should be solvent extracted from the alkali metal halide solution, or a multi-channel type instrument should be used to measure and subtract the background absorbance in the flame. The main channel analyzes the object element, and a sub-channel equipped with a deuterium lamp and using a continuous spectrum light source analyzes the background absorption of the flame. Fig. 4.4 shows the atomic absorption spectroscopy apparatus equipped with a sub-channel. Fig. 4.5 gives an example of the subtraction. First, given amounts of mercury were added to sea water samples and analyzed by the main channel. Then, the background was analyzed by the sub-channel equipped with a continuous spectrum deuterium lamp. The true concentration was obtained by subtracting the data obtained. In the case of atmospheric samples, the possibility of such molar absorption and optical dispersion should be borne in mind, especially when the sample is collected near the coast.

TABLE 4.3. Analytical conditions and sensitivity limits of some elements

Symbol	Element	Analytical line (nm)	Suitable flame	Sensitivity limit†
Ag	Silver	328.1	Air–C_2H_2	0.002
Al	Aluminium	309.3	N_2O–C_2H_2	0.02
Au	Gold	242.8	Air–C_2H_2	0.002
Ba	Barium	553.6	N_2O–C_2H_2	0.05
Be	Beryllium	234.9	Air–C_2H_2	0.001
Ca	Calcium	422.7	N_2O–C_2H_2	0.001
Cd	Cadmium	228.8	Ar–H_2	0.0005
Ce	Cerium	520.0	N_2O–C_2H_2	
Co	Cobalt	240.7	Air–C_2H_2	0.002
Cr	Chromium	357.9	Air–C_2H_2	0.002
Cs	Cesium	852.1	Air–C_2H_2	
Cu	Copper	324.8	Air–C_2H_2	0.001
Fe	Iron	248.3	Air–C_2H_2	0.003
Ge	Germanium	265.2	N_2O–C_2H_2	
Hg	Mercury	253.7		0.0002
K	Potassium	766.5	Air–C_2H_2	0.003
La	Lanthanum	357.4	Air–C_2H_2	
Li	Lithium	670.8	Air–C_2H_2	0.002
Mg	Magnesium	285.2	Air–C_2H_2	0.00005
Mn	Manganese	279.5	Air–C_2H_2	0.001
Mo	Molybdenum	313.3	N_2O–C_2H_2	0.02
Na	Sodium	589.0	Air–C_2H_2	0.001
Nb	Niobium	334.9	N_2O–C_2H_2	
Ni	Nickel	232.0	Air–C_2H_2	0.002
Pb	Lead	217.0	Ar–H_2	0.006
Pt	Platinum	266.0	Air–C_2H_2	0.02
Rb	Rubidium	780.0	Air–C_2H_2	0.005
Sb	Antimony	217.6	Air–C_2H_2	0.1
Si	Silicon	251.6	N_2O–C_2H_2	0.04
Sn	Tin	224.6	Ar–H_2	0.03
Sr	Strontium	460.7	Air–C_2H_2	0.002
Ti	Titanium	364.3	N_2O–C_2H_2	0.05
Ta	Tantalum	271.5	N_2O–C_2H_2	
V	Vanadium	318.4	N_2O–C_2H_2	0.02
W	Tungsten	255.1	N_2O–C_2H_2	0.5
Zn	Zinc	213.9	Ar–H_2	0.0006
Zr	Zirconium	360.1	N_2O–C_2H_2	
D_2	Deuterium discharge tube			

†The sensitivity limit is the concentration of the sample solution at $S/N = 2$ (ppm). The value varies with analytical conditions.

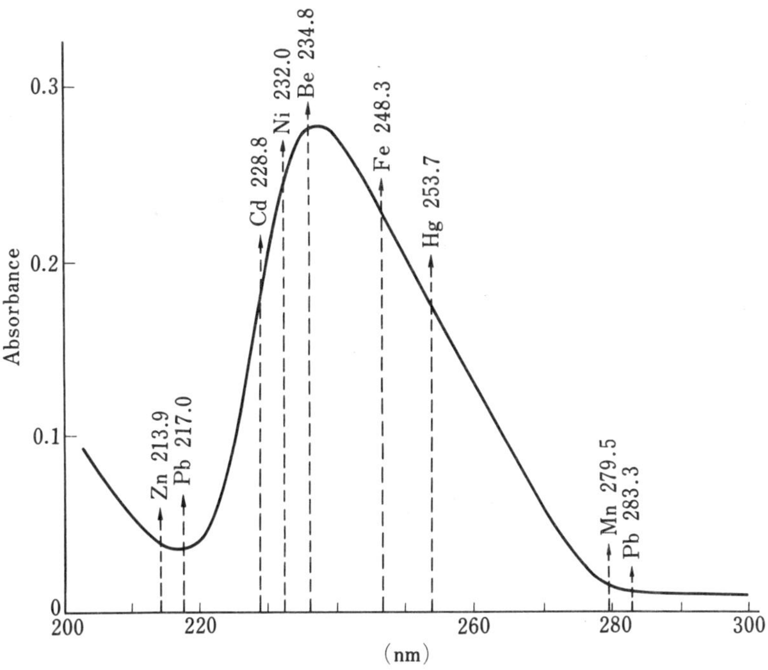

Fig. 4.3. Molar absorption curve of NaCl (10% water solution).

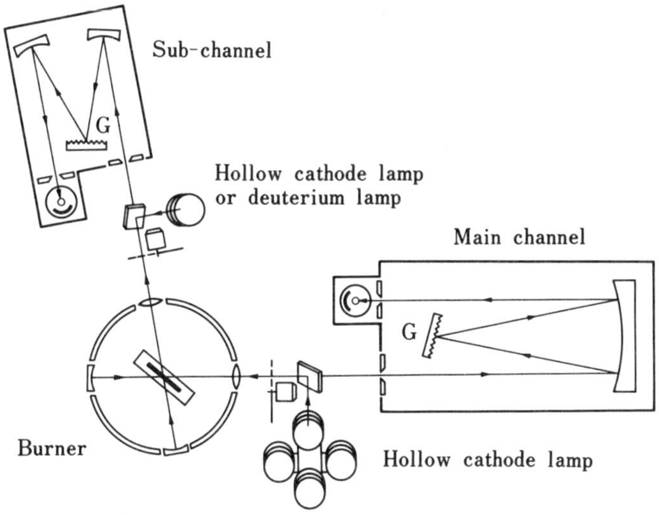

Fig. 4.4. Sub-channel equipped type of atomic absorption spectroscopy apparatus.[6] G = Diffraction grating.

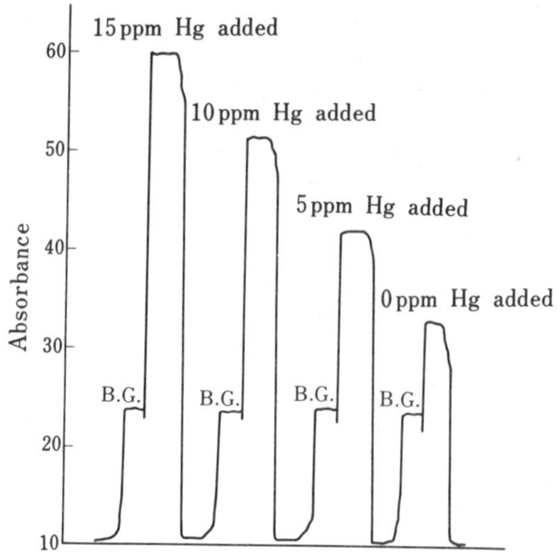

Fig. 4.5. Background due to NaCl. Burner: total consumption type, air-hydrogen flame. Main channel, Hg 253.7 nm; sub-channel, D_2-lamp, 253.7 nm.

ii) Ionizing interference

When the ionizing potential of the element is less than 6 eV, ionization can occur in the flame and this leads to a decrease in the atoms, so affecting the absorbance. Generally speaking, alkaline elements have a low ionizing potential, and ionizing interference can easily occur, especially at high temperatures (nitrous oxide-acetylene flame). It is thus better to use a low temperature flame (air-coal gas) when analyzing for Na or K.

iii) Chemical interference

This derives from incomplete dissociation of the analyzed element. Such a phenomenon easily occurs, especially in a low temperature flame. In this case, a reducing or high-temperature flame is thus preferred. Alkaline earth metals form non-volatile compounds in the flame when Al, Ti, Si or P coexists in the solution, and the absorption potential decreases. Chemical interference is experienced with many other elements such as Mb, Sn, Cr, Mn, and Fe.

c. Selection of flame

Several kinds of flames and their termperature maxima are shown in Table 4.4. The flames of air-acetylene, nitrous oxide-acetylene, and air-hydrogen are the ones commonly used in atomic absorption spectroscopy.

TABLE 4.4. Temperature of various flames

Flame	Max. temp. (°C)
Oxygen—hydrogen	2777
Air—hydrogen	2045
Argon—hydrogen	1577
Air—propane	1725
Oxygen—propane	2900
Air—acetylene	2300
Oxygen—acetylene	3060
N_2O—acetylene	2955

These flames differ as regards temperature, reductivity and flame absorption. The most appropriate flame should be chosen according to the sample and element. The temperature of the flame used in atomic absorption spectroscopy should be sufficiently high as to dissociate the object element and produce free atoms. If the flame temperature is higher than usual, bottom state atoms decrease and excited state atoms increase, the dissociated atoms further ionize, and the overall sensitivity then remains below that with a low-temperature flame. In flame spectrochemical analysis, it is however not sufficient just to dissociate the object element to produce free atoms, but sufficient heat energy should be added to yield many atoms in the excited state. A high temperature flame is thus used. If the flame temperature is too low, the sensitivity decreases. Moreover, interference from coexisting compounds by molar absorption increases. Generally, Cu, Ag, Zn, Cd, Pb, Sn and Se which easily form sulfur compounds, dissociate readily in an air-acetylene flame (2300°C), while Al, Si, V, Mo, Ti, and W which easily form oxides have a strong binding force and require a high temperature flame.

i) Air-acetylene flame

This is the most commonly used of the above three types. Of 68 elements now detectable by atomic absorption spectroscopy, 38 can be detected with an air-acetylene flame.

The defects of this flame are (1) that the temperature is too low for the defection of oxides which do not dissociate easily in the flame, and (2) that flame absorption is quite strong at far UV frequenies, rendering the signal of elements under 230 nm too weak. Sufficient sensitivity is obtained by amplifying the signal but this produces strong noise at the base line.

ii) Nitrous oxide-acetylene flame

This flame is characterized by its high temperature and strong reductive atmosphere, which permit the detection of very low concentrations of non-dissociable compounds that remain undetected with the air-acetylene flame. The elements which can be detected only by the nitrous oxide-acetylene flame are Al, B, Be, Si, Ti, V, W, Er, Eu, Gd, Hf, Ho, Ir, La, Lu, Nb, Nd, Pr, Re, Sc, Ta, Tb, Tm, U, Y, Yb, and Zr. These elements can form oxides (e.g. Al o, VO, TiO, ZrO, etc.) which require a large energy for dissociation, for which the temperature of the air-acetylene flame is too low but the temperature of the nitrous oxide-acetylene flame is sufficient. Moreover, in the nitrous oxide-acetylene flame, these oxides cannot form readily due to the strong reductive atmosphere.

iii) Air-hydrogen and argon-hydrogen flames

Flames using hydrogen as a burning gas, have little absorbance at far UV frequencies. Elements with atomic absorption peaks in this frequency area (Table 4.5) can be analyzed more stably and with less noise than in the case of an air-acetylene flame which has strong flame absorption. The sensitivity so far as the elements in Table 4.5 are concerned, does not differ significantly between the two flames, except for Sn which is detected at 6–8 times higher sensitivity by the air-hydrogen flame. The argon-hydrogen flame has a lower flame absorption than the air-hydrogen flame, especially at 200 nm. (It differs 4 times at 193.7 nm.) The argon-hydrogen flame is preferred for the analysis of As and Se. Argon is used to form the sample mist, and the burning hydrogen gas consumes air gathered from around the flame. The flame temperature is rather low. As, Pb, Cd, Se and Zn, whose oxides have a low dissociation energy, can thus be detected at 1.5–2 times higher sensitivity than with the air-hydrogen flame.

TABLE 4.5. Elements suitable for an air (argon)-hydrogen flame

Element	Analytical line wavelength (nm)
As	193.7
Pb	217.0
Cd	228.8
Se	197.0
Sn	224.6
Zn	213.9

d. Organic solvent extraction and its problems

i) Organic solvent extraction

The objects of organic solvent extraction are (1) elimination of anions and interfering elements, (2) enhancement of sensitivity by the organic solvent effect, and (3) concentration of the sample. Organic solvent extraction may be performed by a combination of ordinary organic solvents. Among the many extracting reagents, DDTC-Na (sodium diethyl dithiocarbamate) and APDC (ammonium pyrrolidine dithiocarbamate) are the most commonly used. Dithizon (Dz) is also used in the analysis of lead and cadmium. These extracting reagents form indissoluble chelated compounds (Table 4.6), and the chelated compounds are extracted to MIBK (methyl isobutyl ketone) or *n*-butyl acetate. The organic phase is directly injected, or back-extracted to a water phase and then injected into the flame, i.e.

TABLE 4.6. Complex formation

IVb	Vb	VIb	VIIb		VIII		Ib	IIb	IIIa	IVa	Va	VIa	
Ti	V	Cr	Mn	Fe	Co	Ni	Cu	Zn	Ca		As	Se	DDTC:
		Mo				Pd	Ag	Cd	In	Sn	Sb	Te	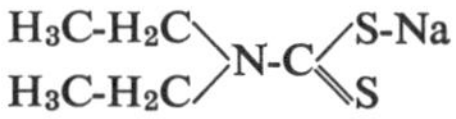 $(H_3C\text{-}H_2C)_2N\text{-}C(\text{-}S\text{-}Na)\text{=}S$
		W						Hg	Tl	Pb	Bi		
		U											
	V	Cr	Mn	Fe	Co	Ni	Cu	Zn			As	Se	APDC:
	Nb	Mo		Ru	Rh	Pd	Ag	Cd		Sn	Sb	Te	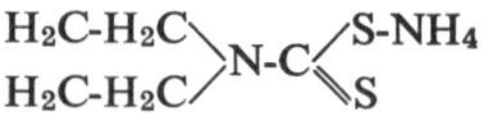 $(H_2C\text{-}H_2C)_2N\text{-}C(\text{-}S\text{-}NH_4)\text{=}S$
		W		Os	Ir	Pt	Au	Hg	Tl	Pb	Bi		
		U											
			Mn	Fe	Co	Ni	Cu	Zn					Dz:
						Pd	Ag	Cd	In	Sn		Te	 $S\text{=}C(NH\text{-}NH\text{-}C_6H_5)(N\text{=}N\text{-}C_6H_5)$
						Pt	Au	Hg	Tl	Pb	Bi	Po	

(1) water phase → organic phase → water phase → injection, or (2) water phase → organic phase → injection.

Back-extraction of ions to a water phase (1) is more reproducible and safer than direct injection of the organic phase (2). However, the latter (2) is much easier to perform and the sensitivity is greatly increased. Due to the poorer reproducibility and safety in (2), special care is required especially when using a premix type burner. APDC is the most widely used extracting reagent, although several disadvantages and problems have been pointed out as follows: (1) APDC is much more expensive than DDTC, (2) APDC often forms precipitates during handling, possibly due to lower stability in the ammonium salt than the sodium salt, and (3) contamination by metals often occurs, (one lot contained 5 ppm iron and 0.5–0.8 ppm lead). The characteristics of the extracting reagents (APDC, DDTC, and Dz) are summarized in Table 4.7.

The ideal requirements for the organic solvent are as follows: insolubility in water, a widely different specific gravity from water, low toxicity,

TABLE 4.7. Comparison of APDC, DDTC and Dz as extracting reagents

APDC	DDTC	Dz
Extractable from acid solution (above pH 4)	Extractable above pH 5	Extractable above pH 9 and at pH 5–6
Many elements can be extracted	Less elements can be extracted than with APDC	Less elements can be extracted than with DDTC
MIBK soln. can be used for analysis High sensitivity	MIBK soln. can be used for analysis High sensitivity	Back extraction to acid solution is necessary from $CHCl_3$ soln. or CCl_4 soln.
If the analysis is performed as MIBK soln., the standard solution should be treated in the same way as the MIBK soln.	If the analysis is performed as MIBK soln., the standard solution should be treated in the same way as the MIBK soln.	In the case of back extraction, the standard solution can be the same grade acidified water soln.
Due to colorless extracting solution, difficult to decide the final extraction point	Due to colorless extracting solution, difficult to decide the final extraction point	Extracting soln. is reversed to green at the end point
Solubility of chelates to MIBK is low, leading to precipitation	Solubility of chelates to MIBK is larger than that of APDC	Solubility of chelates to CCl_4 and $CHCl_3$ is very high, and perfect distribution is obtained
The reagent is unstable and dissociates at high temperatures	Stability is higher than that of APDC	Stability is the same as that of DDTC
Expensive	Relatively inexpensive	Inexpensive
Purity is low Contamination of Fe and Pb is found	High purity is obtainable	High purity is obtainable
Easily contaminated while handling	Easily contaminated while handling	Traditionally used, due to ease of handling

and high inflammability. From this viewpoint, MIBK and *n*-butyl acetate are suitable solvents. However, carbon tetrachloride and chloroform which are non-inflammable, are not so good for direct injection, but back-extraction is performed. The mechanism of the organic solvent effect is not well understood. It is supposed to be related to: (1) viscosity, (2) decrease of surface tension, (3) formation of fine mist, (4) increase of vaporizing velocity, (5) high-temperature formation by burning of solvent, and (6) formation of a reductive atmosphere.

Sensitivity can be increased by solvent extraction, when the ratio of the water phase to the organic solvent phase is increased. In such a case, care must be taken that limitation of solvent amount and percent extraction decreases according to the following formula:

$$\% E = \frac{100\, D}{D + V_a/V_o}, \tag{4.1}$$

where D is the ratio of distribution, V_a the volume of the water phase, and V_o the volume of the organic phase. This formula is expressed graphically in Fig. 4.6.

In some official texts, atomic absorption analysis of atmospheric dusts or dust in waste gases of chimneys, is stated to be performed only by organic solvent extraction. However, there is no reason why all kinds of samples should be analyzed by solvent extraction. When the object element is present as a trace, or interference by other elements is found, solvent extraction is necessary. However, if the object element is abundant

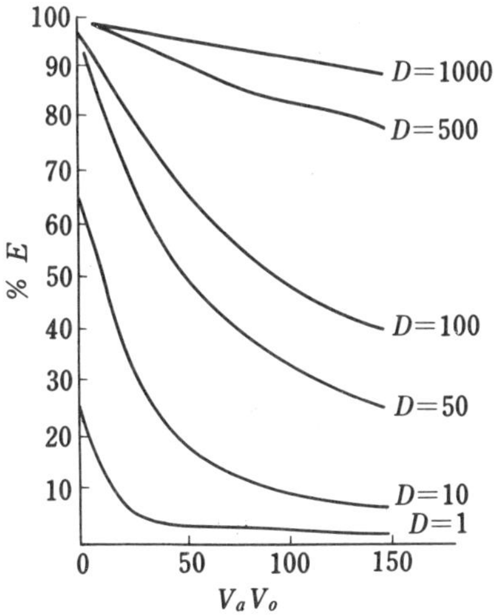

Fig. 4.6. Plots of the volume ratio against percent extraction.

and coexisting elements do not interfere, there is no point in performing solvent extraction. Even if a coexisting element interferes, it can be neglected by drawing calibration curves by the standard adding method, so long as the interference is not by molar absorption or optical dispersion (physical interference). Interference by molar absorption and optical dispersion can be overcome by employing multi-channel type atomic absorption spectrometry.

ii) Problems and key-points of organic solvent extraction

Recovery by solvent extraction is said to reach about 85% when using a separating funnel. However, a series of samples should be handled by the same person, or alternatively, an automatic shaking apparatus should be used. Otherwise, high reproducibility will not be obtained. Differences may also arise in the data obtained when the analyst is a man or woman. The extraction efficiency of some elements changes with the time course after solvent extraction. This is encountered particularly in the case of Mn extraction using an APDC-MIBK combination. Manganese-APDC chelate once extracted to the MIBK phase is back-extracted to a water phase during the time course, since better reproducibility is obtained by back-extraction to the water phase than by direct injection. Direct solvent injection is appropriate to total consumption type burners where the background is stable. Premix type burners display a strong solvent background and instability of the base-line, leading to a decrease in accuracy. Analysts may therefore be tempted to obtain a more stable base-line by decreasing the flow rate of the fuel gas and turning the flame so low as to risk its extinction. Premix type burners are better used with water phase back-extraction than direct solvent injection. Total consumption type burners are easily affected by chemical interference from coexisting elements. Such interfering substances should therefore be separated by solvent extraction, where the solvent can be injected directly into the total comsumption type burner. When the atomic absorption analysis instrument is set up for a prefix type burner, the optical mechanism may be incorporated into the total system. Total consumption type burners may not be acceptable to such instrumentation, so that the sensitivity will be lowered and the lower detection limit increased very much. This means that not every instrument can accept either type of burner. Extreme care should also be paid to losses and contamination during the handling of organic solvents. Great losses may occur, and many of the organic solvents display severe toxic effects on the human body. Adequate precautions should be taken to ensure proper air exchange and the organic solvents should not be handled for long periods. Before attempting to obtain the exact content of the object element by the organic solvent extraction method, sampling and pretreatment of the material should be carried out with high precision. Otherwise, the merits of organic solvent extraction, i.e. increased sensitivity and elimination of interfering ions, may be lost. Sampling errors and errors in preparing the sample solution should be examined and significant figures for the obtained data should always be borne in mind while handling.

Example of organic solvent extraction. Take 100 ml of sample solution into a separating funnel (200 ml). Add 10 ml of ammonium citrate(II) solution (10 w/v %) and a few drops of *m*-cresol purple solution (0.1 w/v %) as indicator. Add ammonia water (1:1) until the solution is slightly purple (pH 7.4–9.0). Add 5 ml of DDTC solution (10 w/v %) and shake. Add exactly 10 ml of *n*-butyl acetate and shake vigorously for one min. Then, allow to settle. After separation of the *n*-butyl acetate phase, evaporate this to dryness with heat. Add exactly 10 ml of nitric acid (1 + 11), dissociate the organic compounds by heat, and evaporate to dryness. Cool, and then add 10 ml of nitric acid (2 + 98). Place the whole solution into a measuring flask (25 ml) and add nitric acid (2 + 98) to the mark line.

e. Key points in the analysis of important elements

Cu. Interference by coexisting elements is small, This is one of the most appropriate elements for atomic absorption spectroscopic analysis. An aqueous solution is injected directly and the problems are generally few. Trace amounts of Cu are extracted with DDTC/MIBK.

Zn. The effects of coexisting substances are not so great, and the sensitivity is high. Organic solvent extraction is needed only in the case of trace amounts. Some glass filters show high blanks.

Pb. The sensitivity is not so high, but this element is appropriate for atomic absorption spectroscopy. Interference is not experienced with an air-acetylene flame, but high concentrations of silicon do have an effect. A low-temperature flame is known to be subject to interference by phosphate and sulfate salts. The analytical lines are 217.0 and 283.3 nm. The former is more sensitive but an air-hydrogen flame should be used in this case. The latter exhibits low noise, and in this case an air-acetylene flame is preferable. Exact checks of molar absorption should be made. Attention should be paid to coexisting Zn when 217.0 nm is used. Extraction is effected with APDC and DDTC, but the former is sometimes contaminated with Pb.

Cd. High sensitivity is obtained. A high concentration of alkali halides (more than 500 ppm) increases the background due to molar sbsorption and optical dispersion. The apparent amount may tend to exceed the true amount. An air-acetylene flame is used, and 228.8 nm is measured

Cr. Many problems are experienced. Sensitivity may be increased by using abundant fuel in an air-acetylene flame, but the background concurrently increases. The apparent amount may become much higher than the true amount. Interference decrease and sensitivity increases when an oxidative flame or nitrous oxide-acetylene flame (red feather) is used. The sample is apt to change to chromic acid (CrO_4^{2-}) during pretreatment. Interference occurs readily with Ca^{2+} or Pb^{2+}, and when these elements coexist in large amounts, the sample is better reduced to the Cr^{3+} form.

Fe. Strong interference occurs with a low-temperature flame, but decreases with a high-temperature flame. Some interference by Si and Ti is experienced with an air-acetylene flame. The analytical line is 248.3 nm, which is close to the molar absorption peak of NaCl. The possibility of this effect must be borne in mind when analyzing for trace amounts of Fe.

f. Calibration curves

i) General calibration method

This method is the most widely used. A series of standard solutions is prepared by dilution of a standard solution containing a given concentration of an element. These solutions are injected and their absorbance is measured. A calibration curve is plotted from the absorbance (ordinate) at each concentration (abscissa). The concentration of an unknown solution is then read off using its measured absorbance value. However, this method does not exclude the effects of interference by coexisting substances, so that the apparent amount of ten differs from the true amount.

ii) Standard adding method

This method consists of adding a series of given amounts of standard solution to the sample solution, and obtaining the absorbance values. A calibration curve is drawn and assumed to be linear to absorbance 0. This point gives the concentration of the object element. Supposing that 5 μg/ml of Mn was contained in the sample solution, and that the apparent amount was 5 μg/ml after one measurement, this still cannot be considered a reliable datum. However, when given amounts of Mn have been added to 1 ml of solution (say 2.5, 5 and 10 μg) and these samples measured to give $5 + 2.5 = 7.5$ μg, $5 + 5 = 10$ μg, and $5 + 10 = 15$ μg, i.e. every sample shows 5 μg more than the added amount, the unknown amount is confirmed as 5 μg. Supposing that the absorbance is proportional to the concentration, the following formula is obtained:

$$A(C) = a(C + x), \tag{4.2}$$

where C is the concentration of the added object element, x the concentration of the object element in the sample, a the proportionality coefficient, and $A(C)$ the absorbance of $(C + x)$.

This formula gives a straight line when the added concentration, C, and absorbance, $A(C)$, are made the abscissa and ordinate. The unknown concentration x is given by the crosspoint on the C axis. Supposing the concentration of Mn in an unknown sample to be x μg/ml, and 5 and 10 μg/ml of Mn to be added to the sample solution to form a series of sample solutions containing x, $(5 + x)$ and $(10 + x)$ μg/ml of the element, then plots of the absorbance values against concentration should give a graph of the type shown in Fig. 4.7. Clearly, the three plots should be on a straight line, and extrapolation of this line to the crosspoint on the abscissa should give the unknown concentration, x.

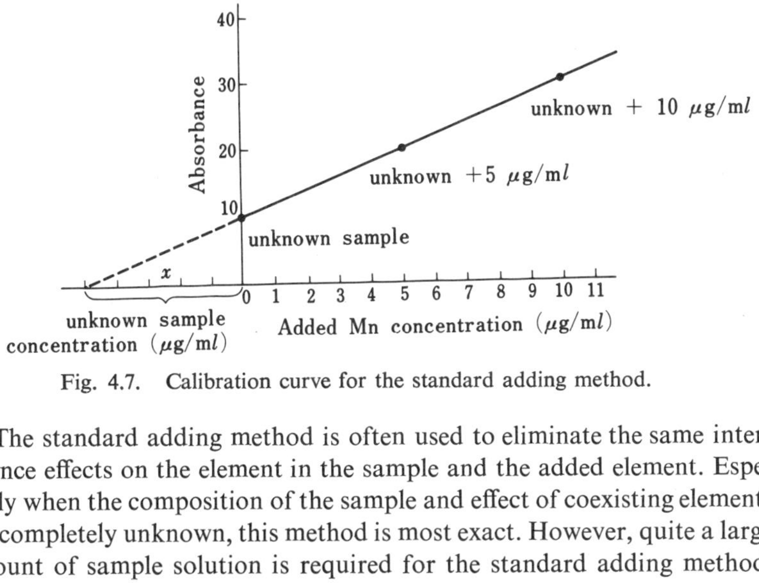

Fig. 4.7. Calibration curve for the standard adding method.

The standard adding method is often used to eliminate the same interference effects on the element in the sample and the added element. Especially when the composition of the sample and effect of coexisting elements are completely unknown, this method is most exact. However, quite a large amount of sample solution is required for the standard adding method, and the unknown sample is diluted several to 10 times. The original solution should therefore be sufficiently concentrated so that the absorbance does not become very low and x difficult to estimate.

ii) Internal standard method

This method is often used in emission spectroscopic analysis. A given amount of internal standard element is added to the sample solution, and the absorbances of the two elements (i.e. the object element and the internal standard element) are measured simultaneously. The ratio of the absorbance of the object element (A_s) to that of the internal standard element (A_r) is obtained. A calibration curve with A_s/A_r as the ordinate and concentration of object element as the abscissa, is then drawn (e.g. Fig. 4.8), from

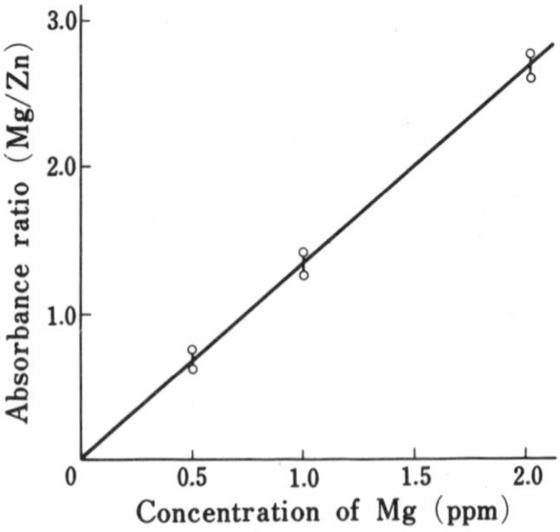

Fig. 4.8. Calibration curve for the internal standard method.

which any sample concentration can be obtained. This method can compensate for the length of light path in the flame, change in the air/fuel ratio, sample viscosity, and change in sucking speed, so providing high overall reproducibility.

Kawasaki *et al.*[6] have enumerated conditions for selecting the internal standard element, as follows: (1) an element not contained in the sample, (2) a congener on the periodic table of elements or an element of close relationship, (3) in the case of a transition element, an element of the same periodicity or of quite close relationship, (4) that the excitation potentials are approximately the same, (5) that the single potentials are approximately the same, i.e. they are quite near elements on the electrochemical series of ionization potentials, (6) that the boiling points or sublimation points are near and the volatilities as compounds are approximately the same, and (7) that the solubilities in the solvent, or reactivities and precipitating abilities with coexisting compounds are similar. The, internal standard element should thus be the closest element to the above conditions, with particular emphasis placed on items (5)–(7).

The internal standard method cannot be applied to all kinds of atomic absorption apparatus: only multi-channel type apparatus is suitable. When the concentration of the object element (in μg/ml) is obtained by such methods, the concentration of a heavy element in the atmosphere is obtained from the following formula:

$$A(\mu g/m^3) = \frac{\text{amount of the element } (\mu g/\text{ml}) \times \text{total vol. of sample (ml)}}{\text{volume of sucked air } (m^3) \times \text{percentage of the collecting area used } \%} \quad (4.3)$$

B. Flameless atomic absorption spectroscopy

Flameless atomic absorption spectroscopy includes (1) the cold vapor method represented by the analysis of mercury, in which $SnCl_2$ is added to the sample solution, the mercury reduced to mercury vapor which is sent into a cell, and the absorption measured, (2) the graphite heated furnace method, in which the sample is heat dissociated in a reductive atmosphere of inactive argon gas, and (3) other methods employing a tantalum or tungsten filament.

a. Analysis of mercury

Three methods are available for collecting mercury from the atmosphere or from some source of environmental contamination, as follows:

(1) trap the contaminant in potassium permanganate/sulfonic acid solution, add $SnCl_2$ to reduce the mercury to vapor, and analyze by flameless atomic absorption spectroscopy;

(2) pass the sample air through an activated carbon column to adsorb the mercury, and then heat and vaporize;

(3) trap with gold or silver to form an amalgam, and then heat and vaporize.

The first method has been used traditionally for the analysis of atmospheric mercury. However, the positive effect of aromatic compounds and ketones is so great that this technique is not applicable to the urban atmosphere or urban sources of contamination. Trapping efficiency is affected by the morphology of the impinger, the pore size, and the morphology of the glass filter at the top of the impinger nozzle.

The principal merits of the second method are (1) that good absorbance of mercury can be obtained, and (2) that the cost of activated carbon is low. However, there is also the demerit that blank mercury differs very much between different makers and different lots. This cannot be easily driven out by heating, and the residual mercury differs according to the heating conditions. This creates great difficulty in preparing mercury-free activated carbon. Activated carbon adsorbs organic compounds and halogen compounds very well, and this may give rise to large positive and negative errors. However, this is not profoundly significant, since one heating is not sufficient to drive out all the mercury. Scaringelli *et al.*[7] have obtained good results by trapping the mercury on an activated carbon column. Their column was heated to vaporize the mercury while passing nitrogen gas through it over silver chips to form silver amalgam. This amalgam was heated and the vapor formed was then analyzed by flameless atomic absorption spectroscopy.

The third method using gold or silver amalgam may emerge as the most

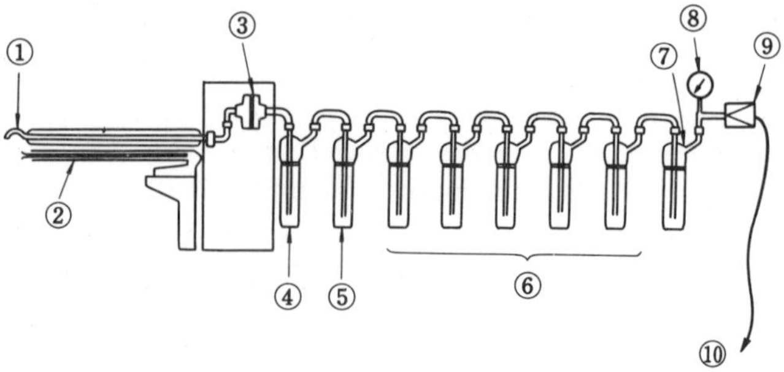

Fig. 4.9. Apparatus for mercury collection. ① Inlet; ② pitot tube; ③ glass fiber filter (heated to 121°C); ④ impinger (containing water); ⑤ impinger (empty); ⑥ gold trap; ⑦ silica gel; ⑧ thermometer; ⑨ flow volume control; ⑩ sucking apparatus (pump, gas meter, orifice).
(Source: ref. 8. Reproduced by kind permission of the American Chemical Society, U.S.A.)

useful method in the future It is not greatly affected by the matrix, both trapping and driving out are easy, and long-term trapping is possible. These are ideal conditions for gathering mercury from the atmosphere or from sources of environmental contamination.

Baldeck *et al.*[8] have reported on the trapping method of mercury from stack gas. The apparatus used is shown in Fig. 4.9. The sample is prepared according to the E.P.A. (U.S.A.) standard method. The glass fiber filter for rejecting dust is maintained at 121°C. The air is then passed through eight impingers. The first two are for eliminating water and the next five are equipped with gold tips at their nozzles to trap mercury. This part of the apparatus is shown in detail in Fig. 4.10. The reason for using the multiple (five) impingers is to keep a high efficiency against oil-contaminated air. After trapping mercury in this manner, the trapping apparatus is heated while nitrogen gas (1 l/min) is passed through it. The mercury vapor is not sent into the cell directly, but once absorbed in potassium permanganate/nitric acid solution. $SnCl_2$ is added to this solution and the mercury is then vaporized by reduction. It is reported that this method has given more than 98% trapping efficiency and is not affected by 6–8% of sulfur oxides.

Oikawa *et al.*[9] have obtained good results for measuring atmospheric mercury by the following method.

Sample trapping. Mercury gas and mercury particles were separately trapped. The method is shown in Fig. 4.11. The sucking flow rate of sample air was 1 ml/min, and mercury particles were removed with a nitrocellulose membrane filter. The interfering substance was then removed by passing the air through a heated furnace (570°C) equipped with a quartz pipe filled with copper oxide and silver chip. The water was eliminated

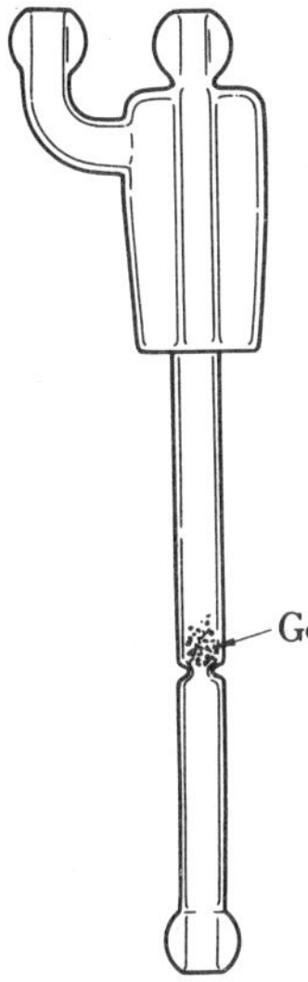

Fig. 4.10. Gold trap of the apparatus in Fig. 4.9.
(Source: ref. 8. Reproduced by kind permission of the American Chemical Society, U.S.A.)

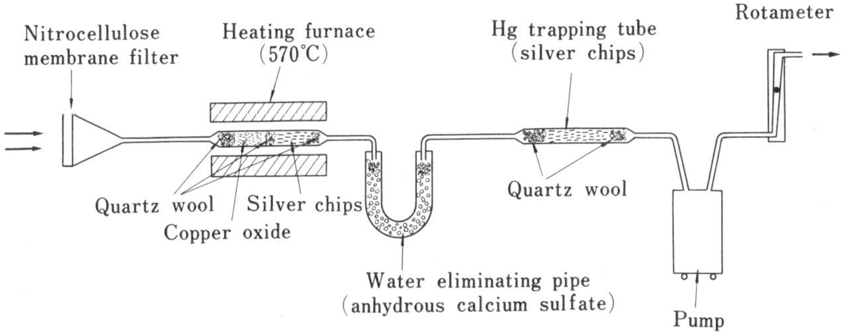

Fig. 4.11. Outline of a sample collecting apparatus.[9]

from this air by passing it through an anhydrous calcium sulfate column, and the mercury gas was trapped as silver amalgam in a trap pipe filled with silver chips. The mercury trap tube was a quartz tube 10 mm ϕ (external diameter) by 7 mm ϕ (internal diameter) by 200 mm long. The silver chip (0.5 mm ϕ) was cut into 3–4 mm lengths, of which 10g was used. The same size of tube was used to remove the interfering substance at the entrance part of which 5g of copper oxide chips (0.3 mm ϕ diameter, length 4 mm) was placed. Another tube was used on the exit side, equipped with 10g of silver chip. The water eliminating U-pipe was filled with anlydrous calcium sulfate.

Method of analysis. The mercury trapped as silver amalgam on the silver chips was driven out by heat vaporization and guided into a flameless atomic absorption spectroscopy apparatus. The analytical line was 253.7 nm. The heat vaporization apparatus is illustrated in Fig. 4.12. The sample trap tube was set in trap heating furnace, which was heated to 680°C. The vaporized mercury was passed through a drying tube filled with magnesium perchlorate at a flow rate of 1 l/min, and then sent into an adsorbing tube. The absorbance was measured using a mercury hollow cathode lamp at a wavelength of 253.7 nm.

The air was cleaned and mercury and other interfering gas eliminated by passage through an activated carbon column. The mercury particles gathered on a nitro-cellulose membrane filter were placed in a combustion boat which was set in a sample dissociating furnace. The vaporized mercury gas was passed through an "interfering substance removing furnace" and drying tube, and then trapped as silver amalgam in the silver fiber trap pipe which was set in the trap heating furnace (at room temperature). The trap heating furnace was then heated to 680°C and the mercury driven out for measurement as mentioned above.

b. Graphite furnace method

A heated Graphite Atomizer or Carbon Rod Atomizer (the name differs according to maker) is used in this method. The apparatus is illustrated in Fig. 4.13 and 4.14.

Method of analysis. To purge, empty heating is carried out under argon

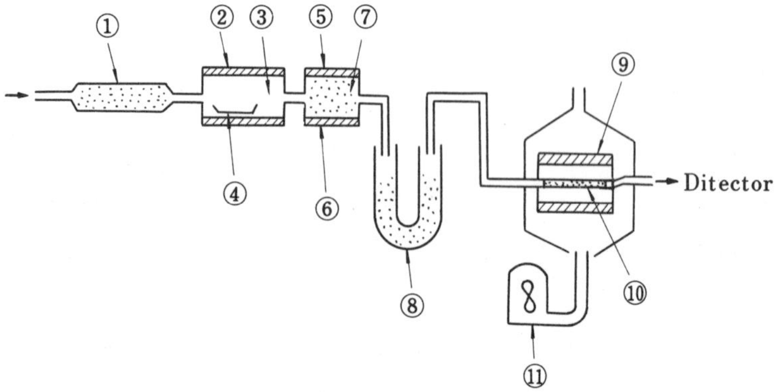

Fig. 4.12. Outline of a heat-vaporization apparatus.[9] ① Activated carbon layer mesh (0.6 g); ② sample dissociating furnace (800°C); ③ combustion catalyst (platinum net); ④ combustion boat; ⑤ interfering substance removing furnace (570°C); ⑥ copper oxide (5 g); ⑦ silver fiber chips (10 g); ⑧ drying tube (25 g magnesium perchlorate); ⑨ trap heating furnace (680°C); ⑩ silver chip trap tube; ⑪ cooling fan.

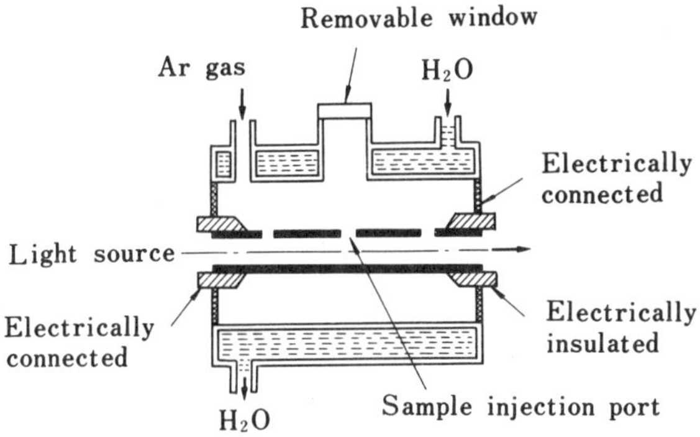

Fig. 4.13. Graphite furnace–I: Perkin-Elmer HGA-70.

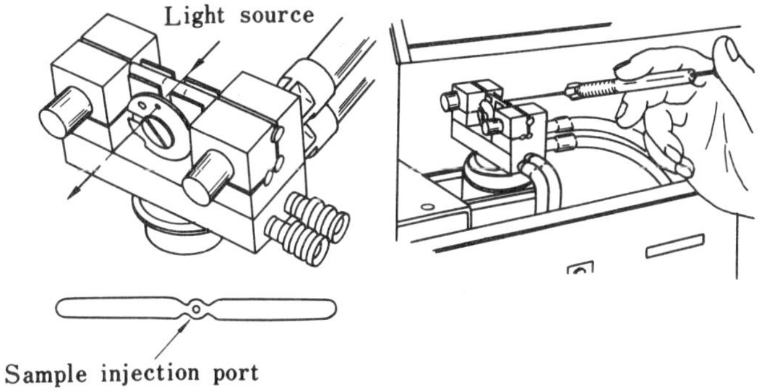

Fig. 4.14. Graphite furnace–II: Varian-Techtron CRA-61.

gas flow, while cooling the apparatus. The removable window is opened and a given amount of sample solution is injected into the cathode tube via a small hole in the graphite tube. The solvent is evaporated to dryness by pre-heating. The voltage is then increased to increase the current. The sample is heat-dissociated with air in the reductive atmosphere over graphite (charring). Co-existing organic compounds are dissociated and removed. Next, the voltage is suddenly increased and the sample atomized at the higher current. Dissociated atoms vaporize simultaneously, and atomic absorption is performed.

In this method, the concentration sensitivity is increased 10 to 100 times compared with the chemical flame method. Also, the sample volume can be as small as μl in order and the detection limit can reach 10^{-10} to 10^{-14} g. One characteristic of the method is that the atomized atoms can remain in the light path for a longer time, since the atmosphere is rather static compared to the dynamic atmosphere of the flame method. This is the reason why the sample volume can be of micro-volume The second characteristic is that the sample is not diluted by gas so that the density of atoms should remain very high. The argon gas is used to prevent oxidation of carbon atoms, which may occur when air is employed. The flameless graphite method is thus highly sensitive, speedy, easy to operate and the sample is atomized in a comparatively static atmosphere. These are its merits compared to the chemical flame method. However, it has been found recently that negative interference occurs when samples rich in salts are analyzed. It must be noted also that the reproducibility is not so good.

Example of analysis of cadmium in the atmosphere with a Heated Graphite Atomizer. There is a need to analyze for atmospheric cadmium, but the concentration is so low that the traditional methods of atomic absorption spectroscopy cannot always give a quantitative result. Janssens *et al.*[10] have recently reported the following method. Atmospheric particles were collected with a cellulose filter and ultrasonicated in nitric acid solution. This sample solution was analyzed by flameless atomic absorption spectroscopy. The cellulose filter was a Watman No. 41, 100 mm in diameter. After passage of 400 m^3 of air through it in 24 h, the filter was placed in a beaker, 50 ml of 0.1 M nitric acid was added and ultrasonication was performed for 10 min. The supernatant liquid was decantated into a 100-ml measuring flask, and the residue was treated with 30 ml of 0.1 M nitric acid. Ultrasonication was performed and the solution was then also placed in the measuring flask which was made up to the marked line. This sample solution was analyzed with a Perkin-Elmer type HGA-70. The characteristic of this method was that washing was carried out only by ultrasonication and not by heating in acid. The results obtained in either methods. The detection limit was 0.04 μg when the injection volume to the graphite tube was 20 μl. The standard deviation of reproducibility was $\pm 4.8\%$.

In addition to this solubilizing method, Brodie *et al.*[11] have reported a direct method by setting the filter in a Carbon Rod Atomizer model 61, attached to a Varian-Techtron AA5. A Millipore filter GSWP (0.22 μm) was washed with hydrochloric acid solution (1 + 2) in order to decrease the blank, and then dried before use. This filter was set in a special graphite sampling cup, 400 ml of sample was passed through it, and then 1000 ppm phosphate (2 μl) was dropped into the graphite cup. The standard solution was treated in the same way. The graphite sampling cup was set in the atomizer, and its contents were dried, chared and atomized. Measurements were made at 228.8 nm. The detection limit was 0.008 μg Cd/m^3 when 200 ml of sample was passed through the filter. No comment was given on reproducibility.

4.4 Emission spectroscopic analysis

A. Characteristics

Emission spectroscopic analysis is a method with utilizes the emission spectrum. When a solid or liquid substance is imparted energy by heating or discharge, the metal elements become atomized at high temperatures. Some part of the atoms shifts to a high energy level, the so-called excited state. When there is a return to a stable low energy level, an emission spectrum characteristic to the element is generated. The emission is detected as a spectrum, and the spectrum is developed on a photo-plate or measured with a photomultiplier tube.

The application of emission spectroscopic analysis to atmospheric pollutants dates from 1958, when the NASN in the U.S.A. utilized it to detect trace metals in the atmosphere. However, although the need for environmental pollutant analysis is increasing more and more these days, emission spectroscopic analysis is tending to lose its position. Instead, atomic absorption analysis is rapidly increasing in the field of metal analysis.

The principal defects of emission spectroscopic analysis may be summarized as follows.

(1) Detection sensitivity is low. The detection limits of atomic absorption analysis and emission spectroscopic analysis are compared in Table 4.8 for the case of 24 h sampling of atmospheric particles (*ca.* 2000 m^3 of air was drawn through). Atomic absorption analysis was always better, except for Va.

(2) Accuracy of analysis is comparatively low. It differs according to the conditions of emission and selection of photo-plate or its developing conditions. A 5–10% standard deviation is considered normal. This means that reproducibility is rather poor, and some elements are known to deviate by as much as 20%. In particular, results for Cr, Fe and the next volatile elements, Cd, Pb, Zn, As and Hg, are not good.

(3) High technical skill is required and individual differences are large, so that a special operator is needed.

(4) The instrument is very bulky and requires considerable operating space.

(5) The instrument is very expensive and the cost per single analysis is

TABLE 4.8. Comparison of the detection limits of atomic absorption spectroscopy and emission spectroscopic analysis (units, $\mu g/m^3$)

Metal	Atomic absorption spectroscopy	Emission spectroscopic analysis
As	0.02	
Ba	0.02	
Ca	0.0002	
Cd	0.0002	0.011
Cr	0.002	0.006
Cu	0.001	0.002
Fe	0.01	0.16
Mn	0.001	0.011
Ni	0.004	0.006
Pb	0.002	0.01
Ti	0.01	—
V	0.01	0.003
Zn	0.0002	0.12

comparatively high, so that is not available to all laboratories undertaking such analysis.

On the other hand, although the sensitivity and accuracy of atomic absorption analysis are good, the number of elements analyzable in a given time is limited. Environmental analysis today requires multi-element analysis, and the greatest characteristic of emission spectroscopic analysis is its ability to perform multi-element simultaneous analysis. Considering the limits to accuracy and sensitivity imposed by arc discharge or spark discharge excitation with photo-plate detection, however, applications to the analysis of metals in the atmosphere remain limited to multi-element qualitative and semi-quantitative analysis. It is nevertheless important to know the approximate number and amounts of contaminant elements in order to improve the accuracy of their quantitative analysis, and such knowledge may give useful information for selecting the suitable method for the object element. Emission spectroscopy shows extremely good sensitivity and quantitativity for the analysis of Be, compared to atomic absorption spectroscopy, X-ray fluorescence analysis and neutron activation analysis. Its continued importance should therefore not be ignored.

B. Methods of analysis

Two methods are employed in the analysis of atmospheric particles by emission spectroscopy. One is to solubilize the sample, and the second is to excite the powder directly. The solubilizing method is represented by the

NASN method described in detail below. This method is to acid extract the sample and place the extracted acid solution in the carbon electrode with a buffer and internal standard. This electrode is then arc- or spark-discharged and the emission spectrum developed on a photo-plate. The density on the photo-plate is measured with a densitometer, and qualitative and quantitative analyses are performed. The analysis may alternatively be done directly using a photomultiplier apparatus. Recently, the sample solution has been directly introduced into an argon plasma torch or plasma jet and measurements done with a photomultiplier apparatus. Such methods can greatly decrease individual differences, and high sensitivity and accuracy are obtained.

In the second, direct powder method, a given amount of sample with an internal standard is placed in the graphite electrode. Following discharge, qualitative or quantitative analysis is carried out with a photomultiplier apparatus.

a. NASN improved method

This method is an improved version of the NASN (U.S.A.) method.

i) Pretreatment of the sample

The sample is collected on a glass fiber filter (8 × 10 in) and 22% of this (2 × 7 in) is ashed in a low-temperature asher.This sample is then transferred to a beaker. (Teflon or Pyrex should be used in order to avoid metal contamination.) Double distilled nitric acid: double distilled water (1:1) solution is added, and heat extraction is performed on a hot plate (110°C) in a draft for 1h. Almost perfect extraction can be attained by double extraction. The extracted solution is filtered through a filter paper (Whatman No. 43 or No. 44), and the filtrate is concentrated to *ca.* 3 ml. A known amount of internal standard solution and the concentrated sample solution are added to a 5 ml measuring flask, which is filled to the marked line with double distilled water. This solution should be kept in a polyethylene bottle in which it will remain stable. Blank solution is obtained by treating the filter paper in the same way as in the case of the sample solution. This pretreatment method presents several problems and a new, improved extraction method is called for.

ii) Preparation of standard solutions

Details for the preparation of standard solutions are shown in Table 4.9. The original solution is placed in a 200 ml measuring flask with a 10 ml hole pipette, and a series of standard solutions is prepared. First, the original water solution is added; then, the original hydrochloric acid solution is added; next, the original nitric acid solution is added. (In cases where a precipitate forms, nitric acid should be added to solubilize it. Special care

TABLE 4.9. Preparation of standard solutions (NASN method)†

Metal	Compound	Original solution (g/100 ml)	Standard solution No. 512 (μg/0.05 ml)	Standard solution No. 256 (μg/0.05 ml)	Standard solution No. 16 (μg/0.05 ml)	Standard solution No. 1 (μg/0.05 ml)	Solvent
Sb	SbO_3	0.040	1.0	0.5	0.031	0.002	HCl
Ba	$Ba(NO_3)_2$	0.320	8.0	4.0	0.25	0.016	H_2O
Be	$BeSO_4 \cdot 4H_2O$	0.020	0.5	0.25	0.016	0.001	HCl
Bi	$Bi(NO_3)_3 \cdot 5H_2O$	0.020	0.5	0.25	0.016	0.001	HNO_3
Cd	$Cd(NO_3)_2 \cdot 4H_2O$	0.020	0.5	0.25	0.016	0.001	HNO_3
Cr	$Cr(NO_3)_3 \cdot 9H_2O$	0.040	1.0	0.5	0.031	0.002	HNO_3
Co	$Co(NO_3)_2 \cdot 6H_2O$	0.020	0.5	0.25	0.016	0.001	H_2O
Cu	$Cu(NO_3)_2 \cdot 3H_2O$	0.400	10.0	5.0	0.31	0.02	NHO_3
Fe	$Fe(NO_3)_3 \cdot 9H_2O$	2.000	50.0	25.0	1.6	0.10	HNO_3
Pb	$Pb(NO_3)_2$	2.000	50.0	25.0	1.6	0.10	H_2O
Mn	$Mn(NO_3)_2$	0.400	10.0	5.0	0.31	0.02	H_2O
Mo	$(NH_4)_6Mo_7O_{24} \cdot 4H_2O$	0.080	2.0	1.0	0.063	0.001	H_2O
Ni	$Ni(NO_3)_2 \cdot 6H_2O$	0.160	4.0	2.0	0.13	0.008	H_2O
Sn	Sn	0.160	4.0	2.0	0.13	0.008	HCl
Ti	$TiCl_4$	0.080	2.0	1.0	0.063	0.004	HCl
V	V_2O_5	0.160	4.0	2.0	0.13	0.008	HCl
Zn	$Zn(NO_3)_2 \cdot 6H_2O$	4.000	100.0	50.0	3.1	0.20	HCl

† The figures are concentrations of metal in solution. Nitrate salts with water of crystallization are strongly hygroscopic and difficult to weigh precisely.

must be taken in winter when precipitation is more likely to occur.) Finally, the flask is filled up to the marked line (200 ml). This gives standard solution No. 512, which is best kept in a Teflon bottle in which it is stable. Standard solution No. 256 is obtained by diluting No. 512 with an equal volume of double distilled water. Standard solution No. 128 is obtained by diluting No. 256 with a further equal volume of water. In this way, a series of standard solutions (No. 512, . . . , No. 4, No. 2, No. 1) is obtained. No. 256 and the other diluted forms are not stable on storage for long periods. The original concentration of each element in No. 512 should be detectable from solutions No. 512 to No. 1, but even after the calibration curve has been prepared, each photo-plate should be checked with No. 32 and No. 16.

iii) Preparation of the electrode and its excitation

The electrode used in the NASN method is a platform type graphite electrode (Hitachi Kasei No. PFT–1901, National AGKSP–L 3948). As shown in Fig. 4.15, it has a constricted neck, which emphasizes the effect of the arc discharge. This electrode is warmed for 10 min at 80°C in a drier, and then coated immediately with 20% paraffin in benzene solution. After a few minutes drying in the drier, 1/3 of the hollow is filled with buffer (LiCl:graphite = 1:2.5 wt/wt) using a Teflon spatula. In the case of measuring the sample solution, a drop of methanol is added and immediately after, 0.05 ml of the sample solution is injected on the electrode with a micropipette. After drying on an "electrode preparation stand", it is once more heat-dried for 1 h at 105°C. In the case of measuring the standard solution, 0.05 ml of the blank extraction solution is injected and then, after drying for a while, 0.05 ml of the standard solution is injected. A drop of methanol is added and the electrode dried on an electrode preparation stand. Heating is for 1 hr at 105°C as in the case of the sample electrode (see Fig. 4.16). The excitation conditions for this method are indicated in Table 4.10. The sample electrode is an anode arranged at an angle of 60° with respect to the counter cathode.

iv) Calibration

The first step is to correspond the spectral details to each element. The analytical lines are shown in Table 4.11. The internal standard is Pd and its analytical line of 324.27 nm is used. Next, the density of each spectral line is measured with a densitometer, and the relative strengths are compared. The metal concentration in the atmosphere is calculated from the sample electrode concentration, C_x (μg/0.05 ml), using the following formula:

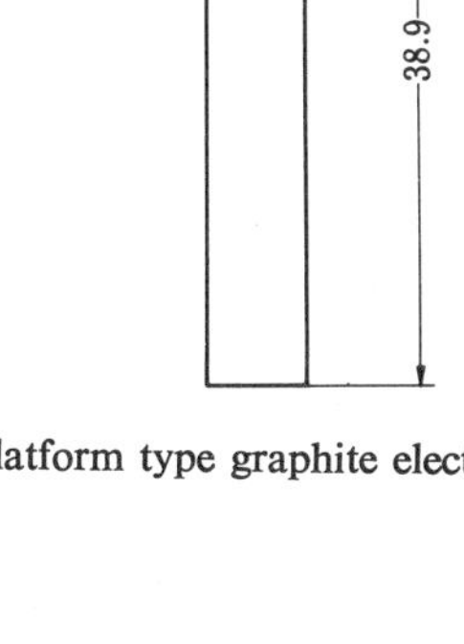

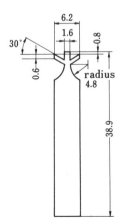

Fig. 4.15. Platform type graphite electrode (units, mm).

$$\text{Metal concn. } (\mu g/m^3) \text{ in atmosphere} = \frac{C_x \times 455}{\text{vol. of air passed through}} \tag{4.4}$$

b. High-speed powder direct analysis (tapping method)

Atmospheric particles are collected on a filter which is set in the electrode directly. After excitation by discharge, the induced emission is measured. This method is speedy and easy to operate. There are a number of reports giving details.[12–13] Particlate substances are concentrated on a

Fig. 4.16. Sample preparation stand and its cover. A heating apparatus is equipped to prevent the deliquescence of LiCl and to dry the solution on the electrode. The "preparation electrode" holding stand consists of an aluminum block with holes, which enables good heat transmission. A plastic cover is used to prevent contamination while handling the electrode and also to protect the human body from volatile acids.

TABLE 4.10. Excitation conditions for quantitative analysis by the NASN method

Excitation source	Baird Atomic Model OB-1
Spectrum dimension	secondary
Voltage	110 V
Current	7 A
Optical system of spectrum	Eagle mounting type
Slit width	25 μm
Electrode width	3 mm
Pre-discharge	3 sec
Exposure time	30 sec
Photo-plate	Kodak SA No. 1
Developer and fixer	Kodak D-19, Kodak fixer
Sector	two step type

TABLE 4.11. Analytical lines in the NASN method

Element	Wavelength (Å)	Element	Wavelength (Å)
Cd	2288.0	Be	2348.6
Fe	2457.6	Pb	2663.2
Cr	2677.2	Cu	3274.0
Sn	2840.0	Sb	2877.9
Mn	2933.1	Ni	3003.6
Bi	3067.7	Mo	3170.3
V	3183.4	Ti	3199.9
Zn	3345.0	Co	3453.5

filter, and the method fully utilizes the characteristics of emission spectroscopic analysis.

The method of Sugimae is introduced here. The electrode is prepared as shown in Fig. 4.17, and after 10 min heating at 80°C, it is coated with paraffin in carbon tetrachloride solution. Buffer (NaF:graphite = 1:4 wt/wt) is then placed on the electrode. The filter is cut out (4 mm in diameter) and placed on the buffer. Next, 0.05 ml of internal standard element, In plus Pd (30 ppm each in 1 N nitric acid solution), is dropped onto it. For the series of standard sample solutions, uncollected filter paper (4 mm in diameter) is placed on the electrode. The series of standard solutions is then dropped onto it. The counter electrode is a graphite electrode, 6 mm in diameter, set at 90° to the main electrode. After exciting these electrodes under the conditions indicated in Table 4.12, the pairs of analytical lines are used for qualitative and quantitative analysis. NaF is employed not only as a buffer but also for decreasing the matrix effect, by dissociating

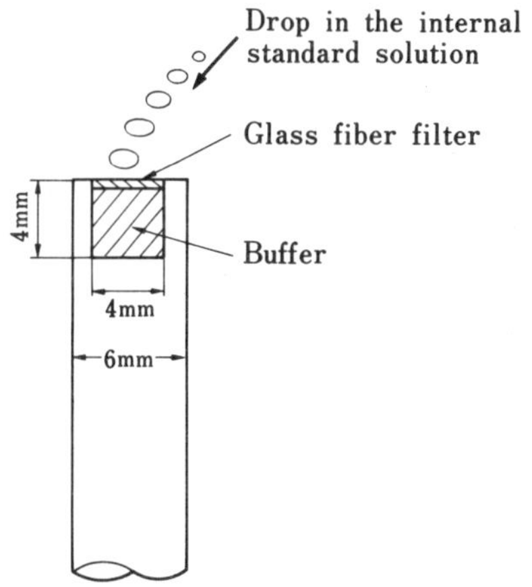

Fig. 4.17. Preparation of the electrode by the tapping method. (after Sugimae)

glass fibers contaminating the filter paper. This increases the accuracy and reproducibility. As the tapping method uses only part of the filter paper, the reproducibility depends upon the state of distribution of the sample. The results of a reproducibility test are shown in Table 4.13. It can be seen that the reproducibility for copper was good, but that for iron and manganese was not. The recovery rate of each element is shown in Table 4.14. Quite good figures, 85–110% were obtained.

TABLE 4.12. Excitation conditions for quantitative analysis by the tapping method (after Sugimae)

Spectroscope	Shimadzu GE-170
Excitation source	Shimadzu Modular Source DC arc
Current	10A
Slit width	15 μm
Electrode width	2 mm
Pre-discharge	none
Exposure time	45 sec
Photo-plate	Fuji Process photo-plate
Developing	FD–131, 20°C, 5 min

TABLE 4.13. Results of reproducibility test (after Sugimae)

Element	1 (μg)	1 (μg)	3 (μg)	4 (μg)	5 (μg)	6 (μg)	7 (μg)	Mean value (μg)	Coefficient of variation (%)
Fe	7.12	6.68	7.59	5.21	5.54	6.81	6.42	6.48	13.1
Mn	0.260	0.266	0.267	0.202	0.225	0.253	0.232	0.243	10.0
Pb	0.824	0.760	0.784	0.815	0.780	0.832	0.872	0.810	4.7
Sn	0.116	0.104	0.097	0.090	0.113	0.107	0.111	0.106	8.6
V	0.073	0.070	0.082	0.074	0.075	0.088	0.083	0.078	8.4
Cu	0.119	0.112	0.122	0.132	0.123	0.115	0.092	0.119	1.2

TABLE 4.14. Results of recovery test (after Sugimae)

Element	Amount added (μg)	Amount detected (μg)	Recovery (%)
Fe	6.50	7.22	111
Mn	0.45	0.38	85
Pb	2.50	2.47	99
Sn	0.25	0.15	106
V	0.30	0.31	103
Cu	0.30	0.31	105

4.5. X-ray analysis

X-ray analysis has been widely employed at the quality control stage of manufacturing processes. Other than this, however, it has been used only by investigators fully familiar with the methodology, or at certain specific laboratories.

Recently, the kinds of environmental contaminants have become very diversified, and it has become necessary not only to measure the density of a certain element by atomic absorption analysis, but also to measure the density of as many elements as possible and to determine their chemical states. X-ray fluorescence analysis, by which both quantitative and qualitative analyses can be made without destroying the specimen, and X-ray diffraction analysis, by which determination of the crystal phase of the metal compounds can be made, have thus been highly estimated and widely used for the analysis of environmental contaminants at various laboratories.

A. X-ray fluorescence analysis

a. Principles and instrumentation

The principle of the equipment used for X-ray fluorescence analysis is illustrated in Fig. 4.18. The secondary beams (fluorescent X-rays) produced by X-ray irradiation become parallel after passage through a collimator, and are then led to a monochromator crystal placed at the center of a goniometer. In this way, the secondary beams are separated into an atomic spectrum. Regarding the separation of the beams, Bragg has given the following formula:

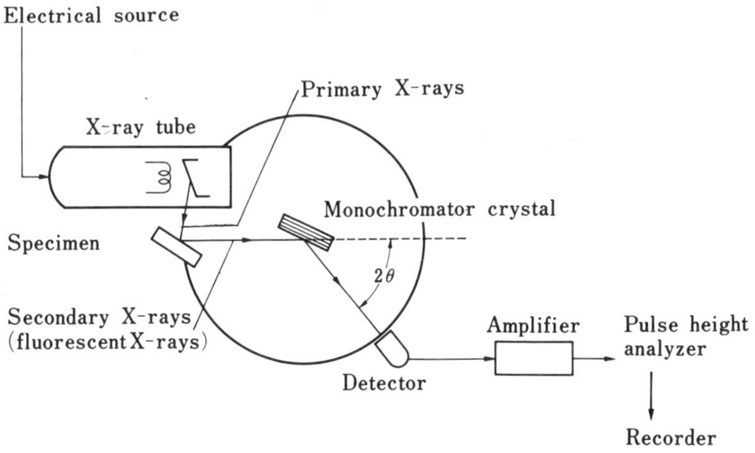

Fig. 4.18. Principle of X-ray fluorescence analysis.

$$2d \sin \theta = n\lambda, \tag{4.5}$$

where λ is the wavelength of the incident X-ray, d the spacing of the lattice planes of the monochromator crystal, θ the angle of incidence, and n = 1, 2, 3, ...

The separated beams are led into a counter tube where the intensities of the X-rays are transformed into electrical signals. These signals are amplified, and then a calculating circuit finally transforms the X-ray intensities into analog or digital quantities.

The equipment consists of three parts, as follows.

X-ray tube. The fixed voltage method is preferable for quantitative analysis. Au, Pt, W, Cr, Ag, Mo, and Rh are used as the counter cathode. Sometimes the characteristic X-rays of the counter cathode overlap with the fluorescent X-rays from the specimen, rendering microanalysis difficult. To avoid such difficulties, an appropriate X-ray tube should be adopted since different tubes give different detection sensitivities. The counter cathodes and elements for which analysis is difficult are related on Table 4.15.

TABLE 4.15. Counter cathodes and elements for which analysis is difficult to perform

Counter cathode	Elements
$_{42}$Mo	Mo
$_{74}$W	Au, Ge, W
$_{78}$Pt	Se, Pt
$_{79}$Au	Se, Au

Monochromator. The monochromator consists of a sample case, collimator or slit, and a monochromator crystal.

Detector. A scintillation counter should be employed for the measurement of X-rays of short wavelength, and a proportional counter using Ar-methane gas for the measurement of X-rays of long wavelength. In the case of multi-element simultaneous analysis, an appropriate detector and monochromator crystal are necessary for each element, respectively.

b. Advantages of X-ray fluorescence analysis

X-ray fluorescence analysis has many advantages, as follows.

(1) The method provides absolute measurements.

(2) Compared with the emission spectrochemical analysis, the peaks are few and readily comprehensible.

(3) The form of the specimen does not affect the measurement. Either liquid, powder or lumps can be used directly.
(4) Qualitative and semiquantitative analyses are easy to perform.
(5) The method is non-destructive, so that the specimen can later be re-analyzed by another method or stored indefinitely.
(6) Elements which have a larger atomic number than that of fluorine are quantitatively analyzable.
(7) Sufficiently strong X-rays give a comparative accuracy of 0.3%. An accuracy of 2–3% is easily attained.
(8) Measurement is rapid and easy to make.

c. Disadvantages of X-ray fluorescence analysis

(1) The equipment is rather expensive, so that ordinary investigational facilities and industrial companies are not easily able to afford it.
(2) The equipment is too large to carry around.
(3) As a rule, the sensitivity is lower than that of other analytical methods such as atomic absorption analysis. The limit of measurement is generally about 0.01 wt%, but can be improved to about 0.001 wt% under optimum conditions of specimen and instrument.

d. Method of quantitative analysis

In quantitative analysis, the following factors are important because of their strong influence on the accuracy of analysis: (1) selection of counter cathode, (2) selection of monochromator crystal, and (3) selection of detector.

For example, the following results were obtained for the analysis of Cd. Analysis with Cd–$K\alpha$ gave poor accuracy due to the high background caused by continuous X-rays. Analysis with Cd–$L\alpha$ and a Cr target X-ray tube gave higher accuracy due to the high excitation efficiency of Cr-K rays. A Rh-target X-ray tube has a poor excitation efficiency and is disturbed by Rh–$L\gamma$, leading to poor sensitivity. By measuring Cd–$L\beta$ with LiF (200), a high intensity X-ray can be obtained, but this is overlapped by the K–$K\alpha$ ray.

The method of X-ray fluorescence analysis of atmospheric Cr, Fe, Pb, V and Ni using DDTC filtercake as a series of standard samples will be described next.

Summary. Sodium acetate solution is added to a solution containing a given amount of Cr, Fe, Pb, V or Ni, and the pH value of the resultant solution is controlled. DDTC solution is added to this solution to effect precipitation. The precipitate is collected with a Millipore filter and made into a filtercake. This filtercake is then dried to form one of the standard samples. In this way, a group of standard samples containing different quantities of the metals is prepared. Cr, Fe, Pb, V and Ni collected from the atmosphere on a Millipore filter are then analyzed by X-ray fluorescence analysis using the calibration curve obtained from the standard samples.

Reagents

(1) Sodium acetate solution: dissolve 300 g of sodium acetate trihydrate (analytical grade) in water and make the volume up to 1000 ml precisely.
(2) DDTC solution: dissolve 2 g of sodium diethyldithiocarbamate trihydrate (analytical grade) in water and make the volume up to 100 ml precisely.
(3) Washing solution: prepare 50 ml of the above-mentioned sodium acetate solution and control the pH value to about 4.2 with hydrochloric acid and aqueous ammonia (1 + 10). Add 30 ml of the DDTC solution to this solution.

(Notes) It is preferable to filter (1) and (2) with a Millipore filter before use. Solutions (2) and (3) should be used within one week of being made.

Original solutions of standard samples

Fe: dissolve 0.100 g of electrolytic iron (over 99.5 % purity) in 50 ml of nitric acid (1 + 4) with heat, and boil out the nitrogen oxide. When the solution has cooled down, add water to make the volume up to 200 ml precisely. This solution contains 0.5 mg/ml of iron.
Ni: dissolve 0.500 g of electrolytic nickel (over 99.8 % purity) in 20 ml of nitric acid (1 + 1). When the solution has cooled down, add 10 ml of sulfuric acid (1 + 1). Heat until white smoke evolves, and drive out the nitric acid. Dissolve the salts which crystallize out in hot water, and when the solution has cooled down, add water to make the volume up to 1000 ml precisely. This solution contains 0.5 mg/ml of nickel.
Cr: dissolve 1.414 g of potassium dichromate in water and make the volume up to 1000 ml precisely. This solution contains 0.5 mg/ml of chromium. (It is preferable to use potassium dichromate that has been crushed in an agate mortar and dried for 3–4 h at 100–110° C.
Pb: dissolve 1.599 g of lead nitrate in water and add 1 ml of nitric acid (1 + 1). Then add water to make the volume up to 1000 ml precisely. This solution contains 1 mg/ml of lead.
V: dissolve 1.148 g of ammonium metavanadate in 200 ml of hot water and make the volume up to 500 ml precisely with water. This solution contains 1 mg/ml of vanadium.

Standard solutions. Prepare the standard solutions by dilution of the original solutions, and use within a short time of preparation.

Method of preparing the filtercake. Add to a 200 ml beaker the quantity of standard heavy metal solution indicated by the figures in Table 4.16. Add water to make the volume up to about 50 ml. Add 10 ml of sodium acetate solution, and after stirring well, add 10 ml of DDTC solution controlling the pH value at 3.9–4.0. Then add water to make the solution up

TABLE 4.16. Quantity of metals (μg) when making filtercakes†

No.	Ni	Pb	Cr	Fe	V
2- 1	5	30	3	60	3
2- 2	5	60	6	120	6
2- 3	5	90	9	180	9
2- 4	5	60	3	120	9
2- 5	5	60	6	180	3
2- 6	6	60	9	60	6
2- 7	10	30	3	120	9
2- 8	10	30	6	180	3
2- 9	10	30	9	60	6
2-10	10	90	3	120	9
2-11	10	90	6	180	3
2-12	10	90	9	60	6

† When the elements to be analyzed are few and the effects of the matrix need not be considered, the general method for making the calibration curve may be employed.

to 100 ml. Stir in an ultrasonic washer for about 1 min and collect the precipitate by suction filtration with a Millipore filter HA (0.45 μm ϕ). The precipitate on the filter wall should be washed with washing solution (5–10 ml). Dry the filter and precipitate with an infrared lamp. Special care should be taken not to contaminate the filter and precipitate in any way.

(Notes)

(1) The pH value after DDTC addition should be adjusted to exactly 3.9–4.0, even if a correction to pH 3.9–4.0 was made before DDTC addition.

(2) The beaker should be placed in the ultrasonic washer. The time for ultrasonication varies according to instrument, but 1–2 min will be appropriate.

(3) When setting the Millipore filter in the funnel, care should be taken that the filter sticks fast to the funnel plate. The suction time should be controlled at 3–5 min in order to prepare a uniform filtercake.

Method of measuring the X-ray intensity. Any kind of X-ray tube can be used but the tube voltage should be kept above 40 KeV. Other conditions of measurement vary between different instruments, so that optimum conditions for the particular instrument in question should be employed.

(Notes)

(1) The kind of sample used tends to show a high background which should be minimized. One way to do this is to set the filter to the sample holder as shown in Fig. 4.19.

(2) The measuring time should be controlled appropriately to keep the statistical error of the net peaks below 2%.

Treatment of data. The abscissa should be marked in μg/cm^2 (obtained

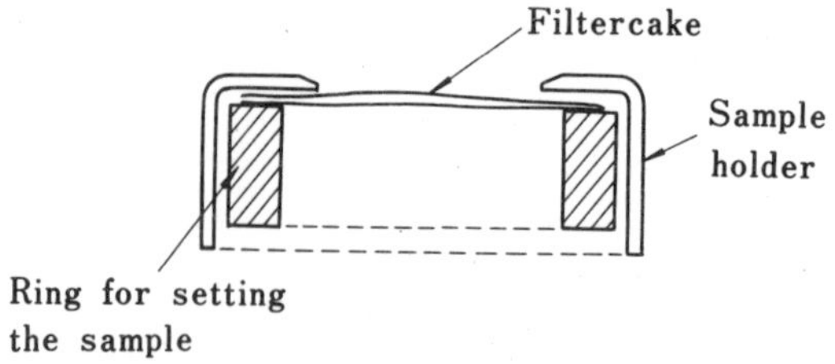

Fig. 4.19. Example of setting the filter to the sample holder.

by dividing the metal content of the filtercake by its filtering area), and the ordinate should be marked according to the intensity of the net peaks in c/s. The calibration curve should be drawn by regression calculation. The confidence interval of the error of an unknown sample is shown to be 99%.

Measurement of the rate of recovery. In the case of analysis with filtercakes as standard samples, more than 99% of the collected metal should be caught in the filtercake. The rate of recovery can be easily measured by the following method. Place the filtercake in a 100 ml beaker and add 20 ml of nitric acid (1 + 10). Dissolve the precipitate on the filtercake by heating. When the solution has cooled down, add water to make the volume up to 100 ml precisely. Then, measure the absorbance(A_{ab}). Treat the filtercake made from the blank (100 ml of water) in the same way and let its absorbance be B_{ab}. Also, collect the standard solution (used to make the filtercake) into a 100 ml beaker. Add 20 ml of nitric acid and water to make the volume up to 100 ml precisely. Measure the absorbance (S_{ab}). Finally, calculate the rate of recovery (G %) from the following formula:

$$G\ (\%) = (A_{ab} - B_{ab}) \times 100/S_{ab} \tag{4.6}$$

If the metal content is only a trace, extract the metal with DDTC-MIBK or dithizon-MIBK, and measure its content by the direct injection method.

e. Energy dispersion X-ray analysis

Energy dispersion X-ray analysis has recently aroused increasing interest due to its high sensitivity, and quick, automatic analysis (using computers). The apparatus for energy dispersion X-ray analysis using a radioisotope as an excitation source, is equipped with an X-ray spectrometer consisting of a Si(Li) semiconductor detector and a multi-channel pulse-height analyzer. Therefore, in contrast to wavelength dispersion X-ray analysis, all of the spectrum can be detected in a short time by this method. The data obtained can be memorized in magnetic tapes and analyzed by computers. In this way, the sensitivity of detection is improved by the order of 10 in some elements, and by the order of 10^2 in others.

This method is the most promising method for X-ray fluorescence analysis since an automatic analysis system can be easily made and combination with neutron activation analysis permits a wide application.

f. Electron microprobe X-ray analyzer (EMPA)

This X-ray microanalyzer is a new scientific tool. The instrument originates from an application of the technique of electron microscopy to X-ray analysis developed by Gunie and Castaing in France in 1949.

The principle of this method is as follows. A finely focused electron beam, 1 μm in diameter, is led to the specimen and the characteristic X-rays from it are separated into a spectrum and analyzed as in the case of X-ray fluorescence analysis. The method aims at the analysis of a micro-spot of the sample or of the biased diffraction of the sample. Recently, it has been employed in the analysis of various substances. This method nevertheless seems very promising for the analysis of atmospheric contaminants. Since the electron beams can be led to the exact, intended spot on the specimen and the occurring characteristic X-rays are separated into a spectrum, both qualitative analysis of the elements and an analysis of chemical shift are possible.

B. X-ray diffraction

a. Demonstration of the need for state analysis

So far, analysis has been performed only on concentrations as metal, i.e. the concentrations of the elements themselves. Little consideration has been given to their actual state in the atmosphere. It is very important, however, to determine the crystal phase and/or co-ordination numbers of the metal components in the air, in view of their different effects on the biological environment, especially on the human body. It is also important both to observe their chemical and physical actions within the atmosphere, in order to elucidate the mechanisms of air pollution and origins of the pollutants, and to determine the state of the metals in the soil as well as the air, together with the sources of contamination.

When the influence of metal components on the human body has previously been discussed, the quantities of metal have generally been taken as the index. However, $MnSO_4$ is said to have about 100 times the toxicity of MnO_2, and α-Fe_2O_3 is much more apt to induce lung diseases than Fe_3O_4. Moreover, it has been confirmed experimentally that α-Fe_2O_3 may serve as a "carrier" of benzo(*a*)pyrene, a carcinogenic hydrocarbon present in polluted air. Then, on its deep penetration into the lung, the incidence of carcinogenesis due to benzo(*a*)phrene is increased.[15) Photo-oxidation and catalytic oxidation have been proposed as the oxidation mechanisms of SO_2 in the air. Compounds of V, Ni, Mn, Fe, etc. which have high chemical activity, are thought to act as catalysts in the oxidation. It has been shown in the experiments of Bracewell and Gall[16] that Fe_2O_3 catalytically oxidizes SO_2 to SO_4^{2-} in moist weather, such as during rain or mist, as follows:

$$2SO_2 + O_2 \xrightarrow{Fe_2O_3 \text{ or } Fe^{3+}} 2SO_3 \tag{4.7}$$

$$Fe_2O_3 + 3SO_3 \xrightarrow{Fe_2O_3 \text{ or } Fe^{3+}} Fe_2(SO_4)_3$$

Standen[17] has shown that MnO_2 reacts with SO_2 to produce soluble $MnSO_4$ and also reacts with NO_2 to produce $Mn(NO_3)_2$.

$$\begin{aligned} MnO_2 + SO_2 &\rightarrow MnSO_4 \\ 2MnO_2 + 3SO_2 &\rightarrow MnS_2O_6 + MnSO_4 \\ MnO_2 + 2NO_2 &\rightarrow Mn(NO_3)_2 \end{aligned} \tag{4.8}$$

Bracewell and Gall reported that even very small amounts of $MnSO_4$ catalytically oxidized H_2SO_3 to H_2SO_4, and compounds of V and Ni also act in the same way.

Oikawa et al.[18] have studied the oxidation of SO_2 by small particles collected on a filter at normal temperatures, and confirmed a highly linear correlation between the quantities of V and Ni on the filter and the quantities of SO_2 oxidized.

b. Principles

When an X-ray beam passes through a crystal, a portion of it is diffracted. The directions and intensities of the diffracted rays vary according to the crystal structure, kinds of atoms composing the crystal, their numbers, and their relative positions. Thus, it is possible to determine the arrangement of the component elements in the crystal, and to say what kind of compound it is.

An X-ray diffractometer consists of an X-ray generator, X-ray tube, goniometer, and counter. Fig. 4.20 gives a schematic drawing of the measuring method. The X-ray beam which occurs at the focus F of the X-ray tube is directed at the specimen, and the reflected beam which is peculiar to the specimen is diffracted in the 2θ direction. By measuring this angle (2θ), it is possible to calculate the spacing of the lattice planes in the crystal, and simultaneously by measuring the intensities of the diffracted rays, the nature and composition of the specimen can be determined. The data are recorded on an automatic recorder or diffraction photographs are taken with an X-ray camera. In order to identify the component elements, the diffraction patterns of the unknown material are compared with those of known ones. Hanawalt's method is widely used as the method for comparison. However, for qualitative analysis, ASTM cards are convenient as a source of data on known materials (Fig. 4.21).

There are about 20,000 ASTM (American Society for Testing Materials)

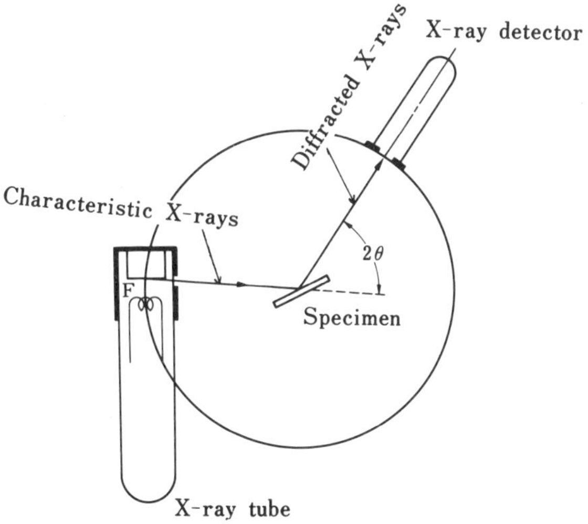

Fig. 4.20. X-ray diffractometer.

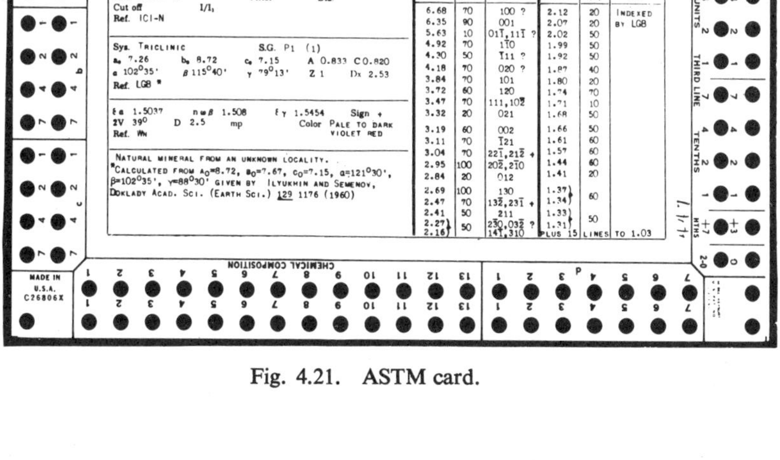

14-426

d	2.95	2.69	6.35	6.68	$Na_4Al_2Si_6O_{17}\cdot H_2O$
I/I₁	100	100	90	70	Sodium Aluminum Silicate Mono Hydrate — Ussingite

Rad. CuK ? λ Filter Dia.
Cut off I/I₁
Ref. ICI-N

Sys. Triclinic S.G. P1 (1)
a₀ 7.26 b₀ 8.72 c₀ 7.15 A 0.833 C 0.820
α 102°35' β 115°40' γ 79°13' Z 1 Dx 2.53
Ref. LGB *

εα 1.5037 nωβ 1.508 εγ 1.5454 Sign +
2V 39° D 2.5 mp Color Pale to dark violet red
Ref. Wn

Natural mineral from an unknown locality.
*Calculated from a₀=8.72, b₀=7.67, c₀=7.15, α=121°30', β=102°35', γ=88°30' given by Ilyukhin and Semenov, Doklady Acad. Sci. (Earth Sci.) 129 1176 (1960)

d Å	I/I₁	hkl	d Å	I/I₁	hkl
6.68	70	100 ?	2.12	20	Indexed by LGB
6.35	90	001	2.07	20	
5.63	10	01$\bar{1}$, 11$\bar{1}$?	2.02	50	
4.92	70	1$\bar{1}$0	1.99	50	
4.30	50	$\bar{1}$11 ?	1.92	50	
4.18	70	020 ?	1.87	40	
3.84	70	101	1.80	20	
3.72	60	120	1.74	70	
3.47	70	111, 10$\bar{2}$	1.71	10	
3.32	20	021	1.68	50	
3.19	60	002	1.66	50	
3.11	70	$\bar{1}$21	1.61	60	
3.04	70	22$\bar{1}$, 21$\bar{2}$ +	1.57	60	
2.95	100	20$\bar{2}$, 2$\bar{1}$0	1.44	60	
2.84	20	012	1.41	20	
2.69	100	130	1.37 }	60	
2.47	70	13$\bar{2}$, 23$\bar{1}$ +	1.34 }		
2.41	50	211	1.33 }	50	
2.27 }	50	2$\bar{3}$0, 03$\bar{2}$?	1.31 }		
2.16 }		14$\bar{1}$, 310	Plus 15	lines	to 1.03

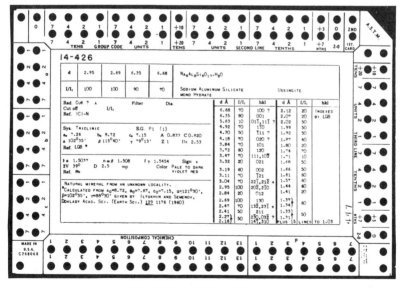

Fig. 4.21. ASTM card.

cards. When using them, attention must be paid to the following points. (1) Among the materials used to prepare the ASTM cards, there are some which have not yet been identified by precise chemical analysis. (2) Many of the values on the ASTM cards were determined by means of diffraction photographs. When using an X-ray counter, therefore, care should be taken over differences in intensities, ability of detection, resolution, etc. (3) Some of the values printed on the ASTM cards may be incorrect.

c. Identification of the crystal phases of atmospheric particles

The crystal phases of materials collected on glassfiber filters with a high-volume sampler have been identified by Oikawa *et al.*[19] by means of X-ray diffractometry. Table 4.17 shows the results, and Fig. 4.22 gives one of the diffraction charts. The sampling locality was Kawasaki City, which is a large industrial area with steel, petrochemical and allied industrial complexes. At each sampling point, α-SiO_2, $NaAlSi_3O_8$, etc. were detected in large quantites, and $CaSO_4 \cdot 2H_2O$ was detected in comparatively large quantities. The former compounds are anticipated to be of soil origin, but the $CaSO \cdot 2H_2O$ may derive from the reaction of atmospheric SO_2 or SO_3 with calcium salts. For example, in urban areas of Kawasaki City where

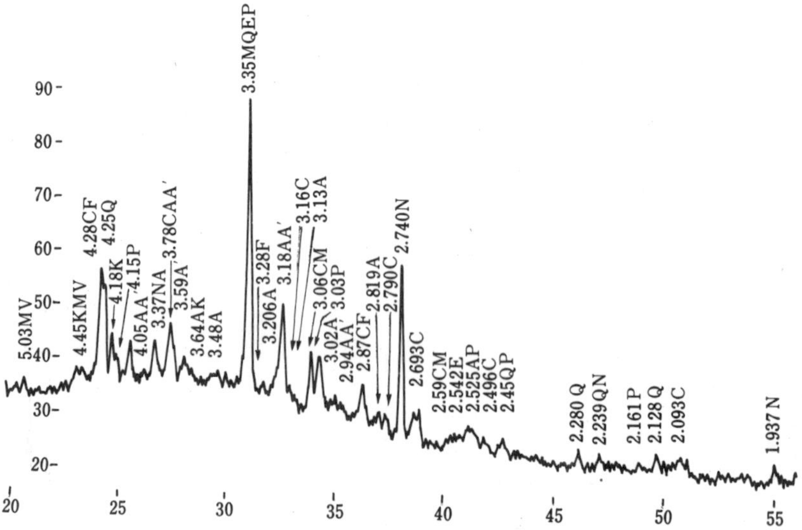

Fig. 4.22. Chart of the X-ray diffractometry of an atmospheric sample collected in the Ohshi area, Kawasaki City (Jan. 1970). Q, α-Quartz (α-SiO_2); N, ammonium chloride (NH_4Cl); C, gypsum ($CaSO_4 \cdot 2H_2O$); A, high temperature albite (1/2 ($Na_2O \cdot Al_2O_3 \cdot 6SiO_2$)); A′, low temperature albite (1/2($Na_2O \cdot Al_2O_3 \cdot 6SiO_2$)); E, endellite ($Al_2O_3 \cdot 2SiO_2 \cdot 4H_2O$); M, muscovite (white mica); K, kaolinite ($Al_2Si_2O_5$ $(OH)_4$); V, vaukisite ($FeAl_2(PO_4)_2(OH_2) \cdot 7H_2O$); F, $(NH_4)_3Fe(SO_4)_3$.

TABLE 4.17. Identification of crystal phases by X-ray diffraction†

	Phase	α-SiO_2				$CaSO_4 \cdot 2H_2O$				$NaAlSi_3O_8$				α-Fe_2O_4				Fe_3O_4				NH_4Cl				$CaCO_3$				NaCl			
	Sampling point / sampling period (1971)	1	2	3	4	1	2	3	4	1	2	3	4	1	2	3	4	1	2	3	4	1	2	3	4	1	2	3	4	1	2	3	4
Winter	9–13 Feb.	●	●	●		·	●	⊙		⊙	⊙	●		·	●	·			·				·	·		⊙	·						
	13–17 Feb.	●	●	●		·	●	⊙		●	·	⊙		·	●	·		·	⊙				⊙			●	⊙	⊙					
	22–25 Feb.	●	⊙	●		●	·	·		⊙	●	⊙		●	·	·		⊙	⊙	·		●				·		·					
	25–27 Feb.	●	●	●		·	●	⊙		⊙	⊙	⊙		·					●			·	⊙			·	⊙						
	27 Feb.–3 Mar.	●	⊙	●		●	⊙	⊙		⊙	⊙	⊙		⊙	●	·		·	·				·	●		·	·						
	3–6 Mar.	●	●	●		●	⊙	·		⊙	⊙	·		⊙	·	·		·				·	⊙			·	⊙	⊙					
Summer	24–27 Aug.	●		●	●	●		⊙	⊙	⊙		⊙	⊙	⊙			·	⊙		·	·									⊙		⊙	⊙
	27–30 Aug.	●		●	●	●		⊙	⊙			⊙	·	●			⊙	⊙			⊙									⊙		⊙	·
	30 Aug.–2 Sept.	●		●	●	⊙		⊙	⊙	⊙		⊙	⊙	⊙		·	·	⊙			·									●		●	●

† ●, Fairly large quantity; ⊙, medium quantity; · , small quantity. 1, Industrial area; 2, industrial area; 3, suburban; 4, urban.

the concentration of sulfurous acid gas is high, a large amount of $CaSO_4 \cdot 2H_2O$ was detected. α-Fe_2O_3 and Fe_3O_4 were encountered in large quantities in the industrial areas, especially at sampling points located near ironworks. It is interesting to note that comparatively large quantities of NH_4Cl and $CaCO_3$ were detected in winter, although little was detected in summer. Abundant NaCl was detected in summer, and appears to derive from salty sea-mists carried by southerly winds and/or winds from the sea.

d. Examination of methods for the quantitative measurement of crystal phases

Only two reports exist on quantitative measurements of crystal phases by means of X-ray diffraction. One is the measurement of amosite and chrysotile by Goodhead *et al.*,[20] and the other is the measurement of α-SiO_2, $CaCO_3$, α-Fe_2O_3, etc. in falling soot by Warner *et al.*[21] However, it seems that neither report has been well appraised, and more precise consideration must be given to the methodology, especially that for sample collection.

A brief account of an examination of the measuring methods made by the author's group[22] is given next.

The additive method and internal standard method are commonly used as measuring methods. At first we considered the additive method. In this case, the accuracy itself is good, but many problems have been pointed out, as follows: (1) the sample materials must be pulverulent bodies, (2) the method is unsuitable for measuring large quantities of materials, (3) pulverulent bodies must be scraped away from the filter, and (4) the absolute quantities can never be measured precisely. We therefore tried the "Direct Method", i.e. internal standard method. Atmospheric particles were collected on membrane filters of nitrocellulose and dissolved in acetone. Using TiO_2 (rutile type) as the internal standard, α-Fe_2O_3 and Fe_3O_4 were measured by the Direct Method. Fig. 4.23 shows the flow chart for this method.

Selection of collection filter. Many examinations were made, chiefly of acetone-treatment of nitro-cellulose and of acetyl-cellulose filters from various makers such as Satrius, Millipore, German, Toyo, etc., and it was found that Toyo-TM80 (made of nitrocellulose) gave good results by acetone treatment.

Selection of internal standard. KCl and TiO_2 (rutile type) were examined. The former reacted on the samples, producing $K_2SO_4 \cdot CaSO_4 \cdot H_2O$. On the other hand, it was confirmed that TiO_2 (rutile type) was adequate enough, since it did not react with the sample materials or with acetone. Subsequently, the pulverization effect on rutile was examined. The relationship between pulverization time (0.5, 1, 2, 5, 10, 20, 30, 60, and 120

Sample

Weighing

X-ray diffraction
qualitative chart

Dissolve sample in
acetone and add TiO_2
as internal standard

Dry up naturally

Specimen for diffraction

X-ray quantitative chart

Quantitative measurement
α-Fe_2O_3
Fe_3O_4

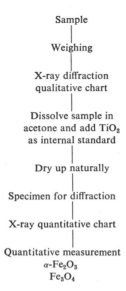

Fig. 4.23. Flow chart for quantitative measurement of crystal phases by the Direct Method.

min) and the X-ray integrated intensities was tested. Although the changes were large up to 1 min, packing densities were high and intensities stable at 2–4 min. Thereafter, the intensities decreased to about 50% at 20 min. We therefore selected a pulverization time of 2–3 min.

Pulverization effects on α-Fe_2O_3 and Fe_3O_4. There was no detectable pulverization effects on α-Fe_2O_3. However, in the case of Fe_3O_4, although there was no change in X-ray integrated intensity up to 7 min, a small decrease in intensity was detected thereafter.

Calibration curves. Fig. 4.24 shows the calibration curves for α-Fe_2O_3 and Fe_3O_4 against TiO_2 (rutile type) as internal standard. The internal standard method can be used only for powdered specimens in principle, so that the ideal case is that where the calibration curve of the specimen before acetone treatment (i.e. the powdered specimen) coincides with that after treatment. Measurements were made 5 times on powdered specimens and acetone-trcated specimens, respectively, repacking them each time. As a result, in the case of the powdered specimens, both the plots for α-Fe_2O_3 and Fe_3O_4 formed straight lines on the calibration curve with a fluctuation width of 4–5%. Similar results were also obtained for the acetone-treated specimens. These calibration curves were therefore considered suitable, and the measurement limit for both specimens was 1 wt%.

Summary

(1) Calibration curves for α-Fe_2O_3 and Fe_3O_4 could be determined

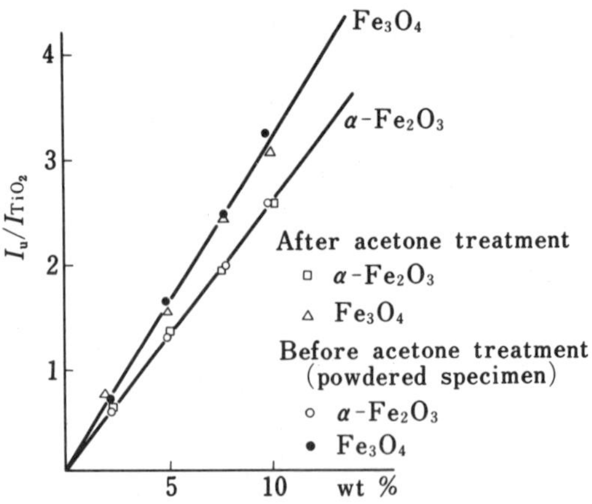

Fig. 4.24. Calibration curves for α-Fe_2O_3 and Fe_3O_4 against TiO_2.

against TiO_2 (rutile type) with a fluctuation width of about 5%. Moreover, the calibration curves for powdered specimens before and after acetone treatment were in good agreement.

(2) The quantities of the aimed phases against all powdered material collected can be determined by this method.

(3) Sampling becomes easier by using an internal standard.

(4) The time required from sample preparation to obtaining results was as short as 2–3 h.

(5) Reproducibility and the *S/N* ratio may be improved by using a graphite-monochromator and rotating sample stand.

4.6. Neutron activation analysis

The application of neutron activation analysis to the field of environmental pollutants is advancing rapidly. The introduction of computers and precise rapid measurements without chemical pre-separation, due to the advent of the high-resolution Ge (Li) semiconductor detector and the 4096 channel γ-ray spectrometer, has rendered instrumental neutron activation analysis a more precise and more rapid analytical technique. It thus has several advantages which no other analytical methods can equal.

Neutron activation analysis can be used for trace analysis due to its high sensitivity for many elements. As a result, the sampling time may be short, and the time-course of pollution can be followed. It is also possible to analyze a large number of elements (*ca.* 40 elements). Previously, the number and kinds of atmospheric metals such as Pb, Cd, Zn, Mn, Fe, Ni, Co, Cu and V, have been determined by atomic absorption spectroscopy, etc., although when considering the total possible effects on the human body, many other injurious metals such as Se, As, Hg, Cr, etc. should be included. Moreover, in the future, even other elements for which no injurious effects have yet been confirmed, may prove to cause serious damage to the human body. It is thus necessary to survey as many elements as possible. Moreover, in this way, the fingerprints of various cities may be taken, and changes in the pollution patterns of the cities can be observed.

The development of studies on the application of instrumental neutron activation analysis to airborne particulate matter has been described by Gordon's group[23)] and by Winchester's group.[24)] In Japan, Hashimoto and the author's group[25–27)], and Mamuro *et al.*[28)] have worked positively in the field. The method has been adopted by the Environment Agency of the Japanese Government for the analysis of samples obtained through the observation system for air pollution in Japan.

A. Principles

Radioactive analysis is performed by counting the radioactivity of a substance which has been made radioactive beforehand. In this case, it is important to know the existing ratio and radiochemical characteristics of the isotopes, due to the differences in their radiochemical natures.

The quantitative relation for an element (an isotope nuclide) that is made radioactive, is given by the following equation:

$$A = N f \sigma S, \tag{4.9}$$

where A is the amount of radioactivity in disintegrations per second (dps), N the number of atoms in the target nuclide, f the neutron flux (n/cm^2.s), σ the thermal neutron cross-section for the nuclear reaction (barn (10^{-24} cm^2)), and S the saturation factor.

Thermal neutrons are most generally used as the activation source in radioactivation analysis. The reason is that they are most reactive and can be obtained stably at high density in a nuclear reactor. The utmost value for the thermal neutron flux obtained in usual nuclear reactors is 10^4 neutrons cm^2. sec.

To circumvent the inconvenience of using a nuclear reactor, neutron sources from a particle accelerator or isotopes may be employed. The values of their thermal neutron flux are, however, 10^6–10^9 and 10^2–10^5 n/cm^2.s, respectively, so that they are unsuitable for trace analysis. The ease of reaction for a given isotope nuclide of an element is measured by the cross-section, σ, in units of 10^{-24} cm^2 (barn). The millibarn is used as an assistant unit.

Radioactive elements begin to disintegrate immediately after they are formed. Therefore, the apparent amount of the product is related to the half-life of the nuclide formed. This relation is expressed as the saturation factor, S, in Eq. (4.9).

$$S = 1 - e^{-\lambda t_i} \tag{4.10}$$

Where λ is the disintegration constant for the radionuclide formed, and t_i the irradiation time (sec).

As the irradiation time becomes longer, the saturation factor increases and approaches a value of 1. The rate of increase in the saturation factor becomes greater as the half-life of the radionuclide formed becomes shorter. The reason is that the disintegration constant (λ) in Eq. (4.10) is related to the half-life in the following way:

$$\lambda = \frac{\ln 2}{T}, \tag{4.11}$$

where T is the half-life (sec).

Short-lived nuclides are formed efficiently by short-time irradiation. However, in order to form long-lived radioactive nuclides, it is not economically favorable to irradiate until the saturation factor approaches unity. The saturation factor has a value of 1/2 for an irradiation time equal to one half-life and a value of 3/4 for an irradiation time of two half-lives. Up to this irradiation time, the efficiency of forming radionuclides is great. Fig. 4.25 illustrates the relationship between irradiation time and rate of radionuclide formation. The radioactivity of the sample, taken from a

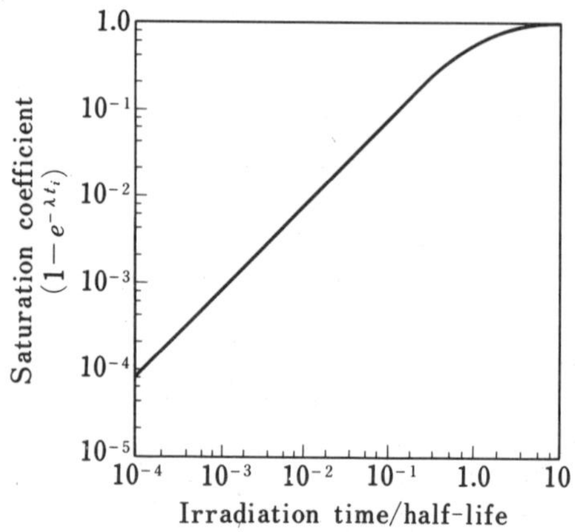

Fig. 4.25. Relationship between irradiation time and saturation coefficient (unit, half-time).

nuclear reactor after irradiation, decays according to its characteristic half-life. Therefore, after the irradiation time and cooling time have passed, the amount of radioactivity of the nuclide is given by the following equation:

$$A = \frac{\omega Av\, a}{100M} f\sigma(1 - e^{-\lambda t_i})e^{-\lambda t_c}, \tag{4.12}$$

where M is the atomic weight of the isotope nuclide, ω the weight of the target element in the sample (g), Av Avogadro's number, a the ratio of the particular isotope nuclide (g), and t_c the cooling time (sec).

From the above equation, it follows that long-lived nuclides form and decay only with difficulty, whereas short-lived nuclides form and decay easily. Thus, in the case of a short-lived nuclide, high analytical sensitivity is obtained by measuring the radioactivity immediately after the short-time irradiation. On the other hand, for a long-lived nuclide, high analytical sensitivity can be obtained by counting for a long time after the decay of the short-lived nuclide. These relationships are illustrated in Fig. 4.26.

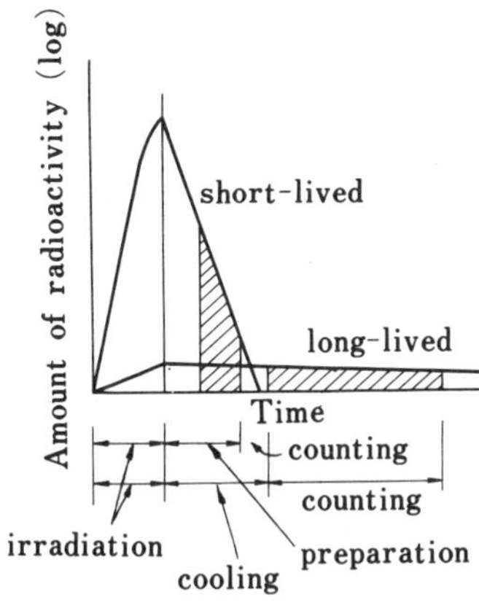

Fig. 4.26. Formation and decay of long-lived and short-lived nuclides. (after Hashimoto)

B. *Analytical procedures*

a. Sample Collection

For sample collection, low volume air samplers of 20–30 1/min suction volume which are used mainly for long successive collection, and high-volume air samplers of 100–500 1/min suction volume, are often employed. When selecting the filter, it is necessary to pay proper attention to the components and quality of the material used. The most generally used filters made of glass fiber are unsuitable for radioactivation analysis. Membrane filters made of nitro-cellulose or cellulose-acetate are generally employed, while polystyrene filters are also commonly used. In particular, polystyrene filters are suitable with high-volume air samplers of 100–500 l/min suction volume.

b. Sample for irradiation

The exclusive sample container for irradiation differs according to the nuclear reactor and irradiation hole used. Here, brief mention is made of the sample itself. In the case of long irradiation, it is necessary to give sufficient consideration to the changes in the sample and container caused by neutron flux and the temperature within the nuclear reactor. Quartz-tubes change in color to purple and polyethylene tubes become brown and brittle. Even when such is not the case, as mentioned above, in the case of sealing the sample into a quartz-tube, etc. it is necessary to carry out the sealing process under reduced pressure. It is inadvisable to irradiate a sample collected on a filter, such as aerosol sample, for a long time by high neutron flux, due to the limit to the strength of the filter. It is usual practice to irradiate the sample sealed doubly in polyethylene sacks and to count it in a new sack instead of the outer sack.

c. Standard materials

In order to perform the analysis by the comparison method, it is necessary to prepare standard materials. For the analysis of 10 elements, 10 standard elements are required, and for 30 elements, 30 standards are required. In general, based on this procedure, it is by no means easy to perform the analysis. That is to say, for the analysis of 30 elements, 30 test tubes must be taken and the solution then sucked out of each tube with a pipette and dropped onto the suction paper. This requires successive dropping of the solutions and drying with an IR lamp. Various attempts have been made to reduce the practical bother. One concerns the preparation of the solution, and Table 4.18 illustrates a successful example. The main point is to develop a combination which does not form a deposit and to treat volatile materials aside, so that they may be handled easily.

d. Irradiation and counting

It is a most important matter for neutron activation analysis, especially for instrumental neutron activation analysis, to decide the length of the irradiation, cooling, and counting times. This can be understood from the relationships shown in Table 4.19. The half-life of the radioactivity of an easily excited nuclide is short, whereas it is difficult to excite a long-lived nuclide. The above point is illustrated in Fig. 4.27 by an actual spectrum, i.e. that counted by a 26 cm³ Ge (Li) detector after 2–6 min, 30–90 min and 2–3 days cooling, following 5 min irradiation of a sample collected on a Millipore filter.

In the irradiation-cooling scheme of Dans *et al.*,[29] 5 min and some hours irradiation is performed and counting is then carried out separating the cooling times into 4 groups.

TABLE 4.18. Groups of standard materials by mixed solution

Short-lived nuclides						
Al	Br	Ca	In†			
Cu	Cl	Na				
Mg	I					
Mn						
Ti						
V						
Long-lived nuclides						
Ba	Hf	Sc	Ni	K		
Co	Ce	Eu	Th	Na		
Cr	Ga	Sm	Ag			
Au	Cs	La	Lu			
As†,	Fe†,	Hg†,	Sb†,	Se†,	Zn†,	W†

† Single solution.

TABLE 4.19. Irradiation, cooling, and counting times for some elements
(1) Short-lived nuclides

Element	Isotope	Half-life	Irradiation time	Cooling time	Counting time	γ-rays used (KeV)
Al	^{28}Al	2.31 min	5 min	3 min	400 sec	1778.9
S	^{37}S	5.05 min	"	"	"	3102.4
Ca	^{49}Ca	8.8 min	"	"	"	3083.0
Ti	^{51}Ti	5.79 min	"	"	"	320.0
V	^{52}V	3.76 min	"	"	"	1434.4
Cu	^{66}Cu	5.1 min	"	"	"	1039.0
Na	^{24}Na	15 h	"	15 min	1000 sec	1368.4;2753.6
Mg	^{27}Mg	9.45 min	"	"	"	1014.1
Cl	^{88}Cl	37.3 min	"	"	"	1642.0;2166.8
Mn	^{56}Mn	2.58 h	"	"	"	846.9;1810.7
Br	^{80}Br	17.6 min	"	"	"	617.0
In	^{116m}In	54 min	"	"	"	417.0;1097.1
I	^{128}I	25 min	"	"	"	442.7

(2) Long-lived nuclides

K	^{42}K	12.52 h	2–5 h	20–30 h	2000 sec	1524.7
Cu	^{64}Cu	12.5 h	"	"	"	511.0
Zn	^{69m}Zn	13.8 h	"	"	"	438.7
Br	^{82}Br	35.9 h	"	"	"	776.6; 619.0; 1043.9
As	^{76}As	26.3 h	"	"	"	657.0;1215.8
Ga	^{72}Ga	14.3 h	"	"	"	630.1;834.1; 1860.4
Sb	^{122}Sb	2.75 days	"	"	"	546.0;692.5
La	^{140}La	40.3 h	"	"	"	486.8;1595.4
Sm	^{153}Sm	47.1 h	"	"	"	103.2
Eu	^{162m}Eu	9.35 h	"	"	"	121.8;963.5
W	^{187}W	24.0 h	"	"	"	479.3;685.7
Au	^{198}Au	2.70 days	"	"	"	411.8
Sc	^{46}Sc	83.9 days	"	20–30 days	4000 sec	889.4;1120.3
Cr	^{51}Cr	27.8 days	"	"	"	320.0
Fe	^{59}Fe	45.1 days	"	"	"	1098.6;1291.5
Co	^{60}Co	5.2 yr	"	"	"	1173.1;1332.4
Ni	^{68}Co	71.3 days	"	"	"	810.3
Zn	^{65}Zn	245 days	"	"	"	1115.4
Se	^{75}Se	121 days	"	"	"	136.0;264.6
Ag	^{110m}Ag	253 days	"	"	"	937.2;1384.0
Sb	^{124}Sb	60.9 days	"	"	"	602.6;1690.7
Ce	^{141}Ce	32.5 days	"	"	"	145.4
Hg	^{203}Hg	46.9 days	"	"	"	279.1
Th	^{233}Pa	27.0 days	"	"	"	311.8

e. Analysis of γ-ray spectra

As the counting equipment for the limited radioactivation analysis mentioned here, a γ-ray detector, peak-height analyzer and recording instrument are used. Scintillators have for a long time been employed as the γ-ray detector. However, at the present time, the recently developed high-resolution semiconductor detector is often used. The difference in the spectra is illustrated in Fig. 4.28, taking aerosol as the sample. The upper spectrum was obtained with a scintillator with a NaI (Tl) crystal, and the lower spectrum with a Ge (Li) semiconductor detector. The high degree of resolution of the semiconductor detector can be clearly seen. The half-height-width for Ge (Li) is 2–3 KeV in energy, whereas that for NaI (Li) is 20–50 KeV. Ge (Li), often called "Geli", is a germanium detector partly

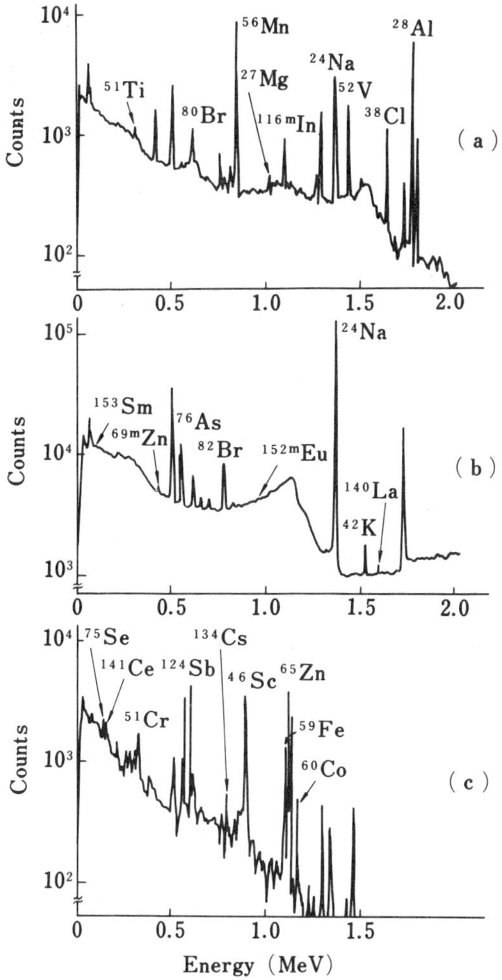

Fig. 4.27. Time course of a γ-ray spectrum based on cooling times of 2–6 min (a), 30–90 min (b), and 2–3 days (c). (after Hashimoto)

diffused with lithium. When the diffusion layer of the lithium spreads, the detector is no longer suitable for use. It is therefore always necessary to cool it with liquid nitrogen.

In order to treat spectra of the type obtained with such a high-resolution detector, a multi-channel spectrometer is required, and a spectrometer with 1000–4000 channels or more is suitable. The spectrum obtained with the Ge (Li) detector involves the radioactive rays from multi-nuclides and is therefore suitable for multi-element analysis. On the other hand, it is dif-

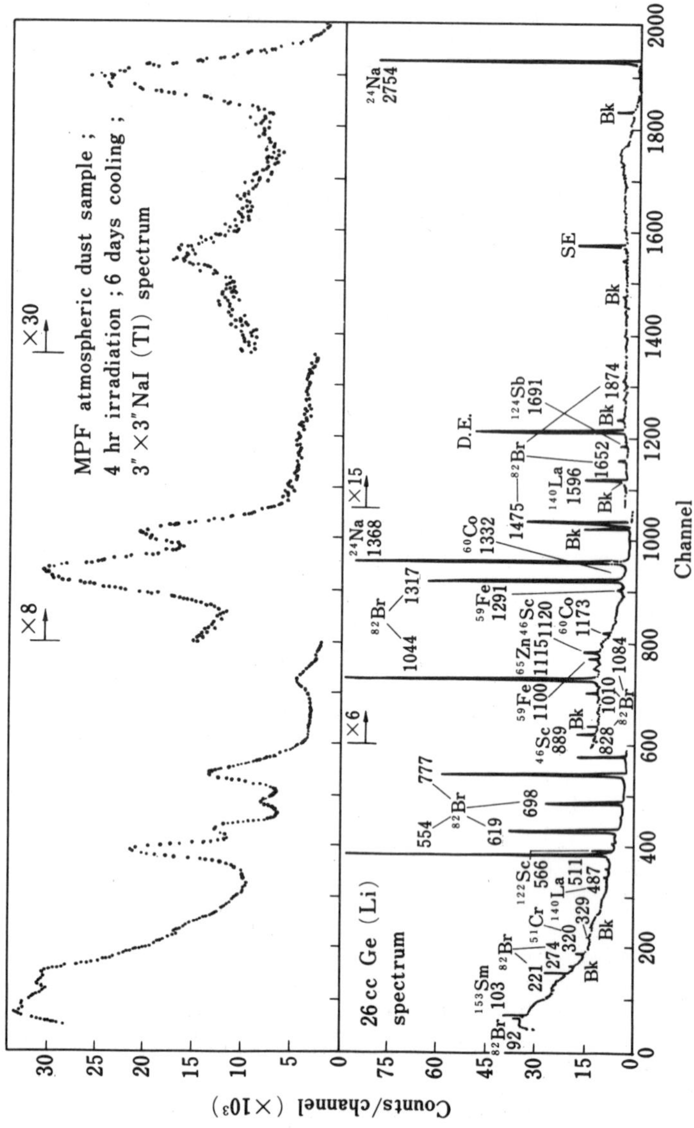

Fig. 4.28. Comparison of γ-ray spectra obtained using NaI (T1) (above), And Ge (Li) (below) (after Hashimoto).

ficult to calculate the peak areas one by one manually. It is usual therefore to use a computer for the analysis of the spectrum. The output of the spectrometer is recorded on magnetic or paper tapes and is then fed into the computer. Spectrometers with an attached minicomputer are also gradually coming into use.

f. Example

The author's group is currently performing the analysis according to the scheme shown in Fig. 4.29, and the data obtained for the γ-ray spectrum are memorized on magnetic tapes to be treated by computers. Examples of analytical data from various localities in Japan are given in Table 4.20, and the approximate levels of sensitivity obtained in the case of using our scheme are shown in Table 4.21.

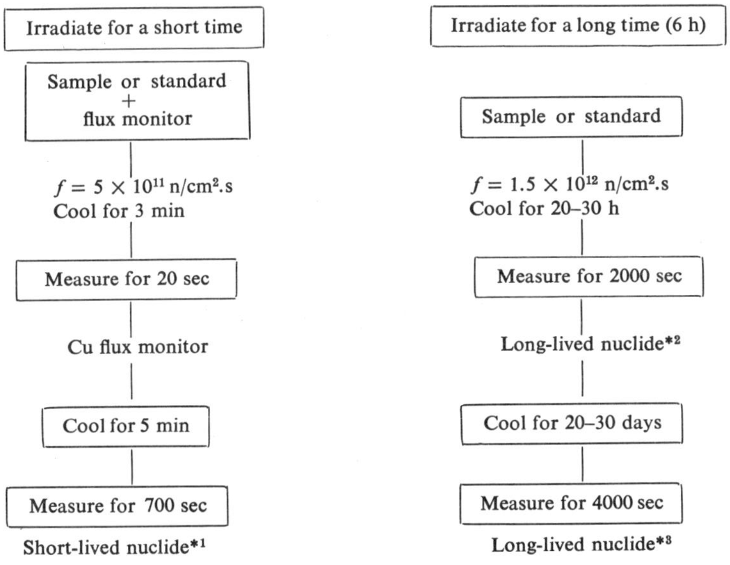

*1 Half-life (T) = several minutes to several tens of hours.
*2 Half-life (T) = several tens of hours to several days.
*3 Half-life (T) = several tens of days to several years.

Fig. 4.29. Analytical scheme for neutron radioactive analysis.

TABLE 4.20. Elemental analysis data at various places in Japan

Measuring point	Harada area, Toyonaka City	Kawasaki City	Shiroyama, Matsuyama City	Hachinohe city office
Sample weight (g)	0.0214	0.020!	0.0109	0.0198
Volume (m^3)	387	321	385	422
Element				
Na	0.64	1.03	0.88	0.65
Mg	0.66		0.86	
Al	0.99	1.04	0.68	1.45
Cl	0.42	0.18		1.0
K	1.04	0.88		
Ca	1.14	1.08	0.74	
Sc	1.4×10^{-4}	1.5×10^{-4}	8×10^{-5}	3.3×10^{-4}
Ti	0.079			
V	0.021	0.080	0.030	0.034
Cr	0 009	0 060	0.004	0.004
Mn	0.090	0.26	0.039	0.031
Fe	1.19	2.23	0.52	0.92
Co	7.4×10^{-4}	0.002	0.0018	5.6×10^{-4}
Ni		0.068		
Cu	0.040			
Zn	0.27	0.71	0.08	—
As	0.016	0.016	0.0050	0.007
Se	0.0018	0.020	0 0013	0.0022
Br	0.034	0.040	0.018	0.043
Rb	0.0027	0.0019		
Ag	8×10^{-4}	0.0024		9×10^{-5}
Cd	0.010	0.030		
In	1.2×10^{-4}	2.4×10^{-4}	1.0×10^{-4}	$<1 \times 10^{-4}$
Sb	0.0097	0.015	0.013	0.0023
Cs	3.8×10^{-4}	3.7×10^{-4}	1.4×10^{-4}	$<2 \times 10^{-4}$
Ba		0.025		0.04
La	6.2×10^{-4}	7.0×10^{-4}	2.0×10^{-4}	5×10^{-4}
Ce	0.0013	0.0016	7.3×10^{-4}	8×10^{-4}
Sm	1.9×10^{-4}	2.0×10^{-4}	1.1×10^{-4}	1.1×10^{-4}
Lu	1.0×10^{-3}	7×10^{-6}		1×10^{-6}
Hf	3.8×10^{-4}	6×10^{-5}		7×10^{-3}
W		0.0025		0.0023
Au				$<1 \times 10^{-5}$
Hg	$<3 \times 10^{-4}$	$<5 \times 10^{-4}$	$<2 \times 10^{-4}$	—
Pb				
Th	1.6×10^{-4}	1.9×10^{-4}	2.0×10^{-4}	3×10^{-4}

TABLE 4.21. Sensitivity of radioactivity analysis to airborne particulate matter

Element	Cooling time	Limit of detection (μg)
Al	3 min	0.04
S	"	25.0
Ca	"	1.0
Ti	"	0.2
V	"	0.001
Cu	"	0.1
Na	15 min	0.2
Mg	"	3.0
Cl	"	0.5
Mn	"	0.003
Br	"	0.02
In	"	0.0002
I	"	0.1
K	20–30 h	0.075
Cu	"	0.05
Zn	"	0.2
Br	"	0.025
As	"	0.04
Ga	"	0.01
Sb	"	0.03
La	"	0.002
Sm	"	0.00005
Eu	"	0.0001
W	"	0.005
Au	"	0.001
Sc	20–30 days	0.003
Cr	"	0.02
Fe	"	1.5
Co	"	0.002
Ni	"	1.5
Zn	"	0.1
Se	"	0.01
Ag	"	0.1
Sb	"	0.08
Ce	"	0.02
Hg	"	0.01
Th	"	0.003

References

1. U.S. Geological Survey Standard–I, Additional Data on Rocks G-1 and W-1 (1965–67); M. Fleischer, *Geochim. Cosmochim. Acta,* **33**, 65 (1969).
2. U. S. Geological Survey Standard–II, First Compilation of Data for the New U. S. G. S. Rocks; F. J. Flanagan, *Geochim. Cosmochim. Acta*, **33**, 81 (1969).
3. N. B. S. Standard Reference Material 1577.
4. N. B. S. Standard Reference Material 1571.
5. Y. Hashimoto, T. Otoshi and K. Oikawa, *Environ. Sci. Technol.*, **10**, 815 (1976).
6. K. Kawasaki *et al.*, *Bunko Kenkyu* (Japanese), **19**, 165 (1970).
7. F. P. Scaringelli *et al.*, *Anal. Chem.*, **46**, 278 (1974).
8. C. M. Baldeck *et al.*, *ibid.*, **46**, 1500 (1974).
9. K. Oikawa *et al., Japan. J. Pub. Health,* **23** 659 (1976).
10. M. Janssens *et al.*, *Anal. Chim. Acta*, **70**, 25 (1974).
11. K. G. Brodie *et al.*, *ibid.*, **69**, 200 (1974).
12. D. W. Lander, R. L. Steiner, D. H. Andersen and R. L. Dahm, *Appl. Spectr.*, **25**, 270 (1971).
13. T. Hasegawa and A. Sugimae, *Bunseki Kagaku* (Japanese), **20**, 840 (1971).
14. T. Hasegawa and A. Sugimae, *ibid.*, **20**, 1406 (1971).
15. U. Saffiotti *et al.*, *Cancer Res.*, **28** (1), 104 (1968).
16. J. M. Bracewell and D. Gall, *Air Pollution*, Proc. Symp. Physicochemical Transformation of Sulphur Compounds in the Atmosphere and the Formation of Acid Smogs, Mainz, W. Germany (1967).
17. A. Standen, *Manganese and Manganese Compounds*, in Kir K-Othmer, *Encyclopedia of Chemical Technology*, 2nd ed., Interscience (1967).
18. T. Okita, K. Oikawa and Y. Ihara, *Koshueiseiin Kenkyuhokoku* (Japanese), **19** (4), 252 (1970).
19. K. Oikawa, T. Iwai, H. Maruyama and A. Murase, *Taikiosen Kenkyu* (Japanese), **6** (1), 85 (1971).
20. K. Goodhead *et al.*, *Analyst*, **94**, 1124 (1969).
21. P. O. Warner *et al.*, *J. Air Pollution Contr. Ass.*, **22**, 887 (1972).
22. K. Oikawa *et al.*, *Bunseki Kagaku* (Japanese), **25**, 524 (1976).
23. W. H. Zoller and G. E. Gordon, *Anal. Chem.*, **42**, 257 (1970).
24. R. Dams, J. A. Robbins, K. A. Rahn and J. W. Winchester, *ibid.*, **42**, 861 (1960).
25. H. Kishi, K. Oikawa and Y. Hashimoto, *Bunseki Kagaku* (Japanese), **25**, 519 (1976).
26. T. Otoshi, T. Shiomi, K. Tomura and Y. Hashimoto,*ibid.* , **25**, 620 (1976).
27. Y. Hashimoto and T. Oe, *ibid.*, **25**, 488 (1976).
28. T. Mamuro, Y. Matsuda, A. Mizohata, T. Takeuchi and A. Fujita, *Radioisotopes*, **20**, 111 (1974).
29. R. Dams *et al.*, *Environ. Sci. Technol.*, **6**, 441 (1972).

INDEX

W

X, Z